Forschungsberichte aus dem Lehrstuhl für Regelungssysteme

Technische Universität Kaiserslautern

Band 5

Forschungsberichte aus dem Lehrstuhl für Regelungssysteme

Technische Universität Kaiserslautern

Band 5

Herausgeber:

Prof. Dr. Steven Liu

Daniel Görges

Optimal Control of Switched Systems

with Application to Networked Embedded Control Systems

Logos Verlag Berlin

Forschungsberichte aus dem Lehrstuhl für Regelungssysteme
Technische Universität Kaiserslautern

Herausgegeben von
Univ.-Prof. Dr.-Ing. Steven Liu
Lehrstuhl für Regelungssysteme
Technische Universität Kaiserslautern
Erwin-Schrödinger-Str. 12/332
D-67663 Kaiserslautern
E-Mail: sliu@eit.uni-kl.de

Bibliografische Information der Deutschen Nationalbibliothek

Die Deutsche Nationalbibliothek verzeichnet diese Publikation in der Deutschen Nationalbibliografie; detaillierte bibliografische Daten sind im Internet über http://dnb.d-nb.de abrufbar.

ISBN 978-3-8325-3096-9
ISSN 2190-7897

Logos Verlag Berlin GmbH
Comeniushof, Gubener Str. 47,
10243 Berlin
Tel.: +49 (0)30 / 42 85 10 90
Fax: +49 (0)30 / 42 85 10 92
http://www.logos-verlag.de

Optimal Control of Switched Systems

with Application to Networked Embedded Control Systems

Optimale Regelung geschalteter Systeme

mit Anwendung für vernetzte eingebettete Regelungssysteme

Vom Fachbereich Elektrotechnik und Informationstechnik

der Technischen Universität Kaiserslautern

zur Verleihung des akademischen Grades

Doktor der Ingenieurwissenschaften (Dr.-Ing.)

genehmigte Dissertation

von

Dipl.-Ing. Daniel Görges

geboren in Koblenz

D 386

Tag der mündlichen Prüfung:	08.07.2011
Dekan des Fachbereichs:	Prof. Dr.-Ing. habil. Norbert Wehn
Vorsitzender der Prüfungskommission:	Prof. Dr.-Ing. Dr. rer. nat. habil. Wolfgang Kunz
1. Berichterstatter:	Prof. Dr.-Ing. Steven Liu
2. Berichterstatter:	Prof. Dr. Karl Henrik Johansson

Contents

Preface

Motivation

Control systems more and more evolve to cyber-physical systems. Such systems possess tight interactions between computation, communication, control and the physical environment [Lee08, Wol09, Poo10, BG11]. Various technical systems already incorporate computation, communication and control, ranging from transportation (automobiles, trains, aircrafts, etc.) over industrial applications (manufacturing and process control) to infrastructure systems (power systems, building automation, etc.). Cyber-physical systems are thus all around us. The technological, economical and sociological impact is consequently tremendous.

Although cyber-physical systems already penetrate practical applications, theoretical foundations are still missing. Methods for modeling, analysis and design of cyber-physical systems accounting for the interactions between computation, communication, control and the physical environment are strongly required. Otherwise, functionality, efficiency, reliability and safety are difficult to guarantee in view of rising complexity. Disciplines like mathematics, control theory, computer science and communications have traditionally developed mostly independently. These disciplines must be conjoined to promote the theoretical foundations of cyber-physical systems.

From a control perspective the following research directions are of particular importance:

First, methods for modeling, analysis and control of systems possessing continuous dynamics stemming from physics and control as well as discrete dynamics stemming from computation and communication must be explored. Such systems are denoted as hybrid systems. Hybrid systems are also relevant beyond cyber-physical systems and thus deserve an investigation in their own right.

Second, methods for modeling, analysis and design of networked embedded control systems must be investigated. Networked embedded control systems are control systems where controllers, actuators and sensors are connected via a communication network and where controllers are implemented on processors being embedded into the application. They thus form main building blocks of cyber-physical systems. Networked embedded control systems are often realized with severely limited computation and communication resources. Integrating scheduling, i.e. the temporal assignment of resources to tasks, and control is therefore of paramount importance to efficiently utilize the computation and communication resources. Corresponding approaches are subsumed under control and scheduling codesign. Networked embedded control systems are furthermore usually subject to computation and communication imperfections like latencies and congestion. These imperfections must be respected during the stability analysis and the control design to ensure stability and performance. Corresponding approaches are subsumed under implementation-aware control.

This thesis aims to add knowledge to both research directions.

Objectives

The first objective of this thesis consists in developing a general framework for optimal control and scheduling of switched systems, an important class and abstraction of hybrid systems. Switched systems consist of a set of subsystems and a switching law that defines switching between these subsystems. Optimal control and scheduling of switched systems amounts to jointly designing a control law and a switching law such that a cost function is minimized. This optimization problem has exponential complexity. Taming the complexity is a major challenge. While stability and stabilizing control of switched systems have been intensely studied in recent years, optimal control and scheduling of switched systems has been rarely addressed. This thesis aims at filling this gap, focusing on discrete-time switched linear systems.

The second objective of this thesis consists in developing a general framework for optimal control and scheduling of networked embedded control systems. Concepts for control and scheduling codesign are sought which ensure stability and allow for balancing between complexity, both offline and online, and performance.

Outline

The thesis is divided into two parts.

Part I focuses on *optimal control and scheduling of discrete-time switched linear systems.*

Chapter 1 gives an *overview* on hybrid systems (Section 1.1), switched systems (Section 1.2), stability and stabilization of switched systems (Section 1.3) and optimal control of switched systems (Section 1.4). Basic concepts are reviewed, related work is discussed and the contributions of this thesis are outlined.

Chapter 2 concerns *finite-horizon control and scheduling (FHCS).* Important concepts used throughout the thesis are presented, including dynamic programming (Section 2.2), complexity reduction via pruning (Section 2.3) and relaxed dynamic programming (Section 2.4).

Chapter 3 addresses *receding-horizon control and scheduling (RHCS)*. An optimization problem based on the receding horizon principle is formulated first (Section 3.1); then the solution to this optimization problem is derived based on the concepts introduced in Chapter 2 (Section 3.2). It is shown that the solution can be expressed explicitly as a piecewise linear (PWL) state feedback control law defined over regions implied by quadratic forms. Closed-loop stability is not guaranteed inherently under the PWL state feedback control law. Therefore, an a priori stability condition based on a terminal weighting matrix and several a posteriori stability criteria based on constructing a piecewise quadratic (PWQ) Lyapunov function and on utilizing the value function resulting from relaxed dynamic programming as a candidate Lyapunov function are proposed (Section 3.3). Furthermore, a region-reachability criterion is deduced (Section 3.4).

Chapter 4 addresses *periodic control and scheduling (PCS)*. Periodic control and offline scheduling ($\mathrm{PCS}_{\mathrm{off}}$, Section 4.1) and periodic control and online scheduling ($\mathrm{PCS}_{\mathrm{on}}$, Section 4.2) are studied. Optimization problems based on periodic switching are formulated (Sections 4.1.1 and 4.2.1) and periodic control theory is reviewed (Section 4.1.2) first; then the solutions to these optimization problems are derived based on periodic control theory and exhaustive search (Sections 4.1.3 and 4.2.2). It is shown that closed-loop stability is guaranteed inherently for both $\mathrm{PCS}_{\mathrm{off}}$ and $\mathrm{PCS}_{\mathrm{on}}$ (Section 4.2.3) and that the solution for $\mathrm{PCS}_{\mathrm{on}}$ can again be expressed as a PWL state feedback control law. Furthermore, several methods for reducing the online complexity under $\mathrm{PCS}_{\mathrm{on}}$ based on relaxation and heuristics are proposed (Section 4.2.4).

Chapter 5 provides *conclusions* and suggestions for *future work*.

Part II focuses on *optimal control and scheduling of networked embedded control systems*.

Chapter 6 gives an *overview* on networked embedded control systems (Section 6.1), implementation-aware control (Section 6.2) and control and scheduling codesign (Section 6.3). Particularly, criteria for classifying control and scheduling codesign methods are elaborated. Various existing methods are classified based on these criteria and the contributions of this thesis are outlined.

Chapter 7 concerns *modeling of networked embedded control systems (NECSs)*. An NECS architecture comprising multiple plants controlled over a single network via a single processor is introduced (Section 7.1). For this NECS architecture a computation and communication model (Section 7.2) and an NECS model (Section 7.3) integrating control and scheduling are devised. A cost function respecting scheduling effects is furthermore proposed to evaluate the performance of the NECS (Section 7.4).

Chapter 8 addresses *control and scheduling codesign of NECSs*. A control and scheduling codesign problem is formulated (Section 8.1) based on the NECS model and cost function given in Chapter 7. It is shown that the NECS model corresponds to a block-diagonal discrete-time switched linear system. The methods for optimal control and scheduling of discrete-time switched linear systems presented in Part I are thus applicable for control and scheduling codesign of NECSs. Partially, the methods must be extended due to the structural properties of the NECS model (Section 8.2).

Chapter 9 addresses *N'-step receding-horizon control and scheduling (N'-step RHCS)*, an extension of RHCS introduced in Chapter 3 for NECSs.

Chapter 10 contains a *case study* on networked control of three inverted pendulums to demonstrate, evaluate and compare the methods presented in Parts I and II.

Chapter 11 provides *conclusions* and suggestions for *future work*.

Publications

Several articles on optimal control and scheduling of discrete-time switched linear systems as well as implementation-aware control and control and scheduling codesign of networked embedded control systems have been published during the doctoral studies. A chronological list of these articles together with their topic and their relation to this thesis is given below.

[IGL07] Michal Izák, Daniel Görges, and Steven Liu. On stability and control of systems with time-varying sampling period and time delay. In *Proceedings of the 7th IFAC Symposium on Nonlinear Control Systems*, pages 1056–1061, 2007. (Implementation-Aware Control)

[GIL07] Daniel Görges, Michal Izák, and Steven Liu. Optimal control of systems with resource constraints. In *Proceedings of the 46th IEEE Conference on Decision and Control*, pages 1070–1075, 2007. (Optimal Control of Switched Systems, Chapter 4, and Control and Scheduling Codesign, Chapters 7, 8, 10)

[IGL08] Michal Izák, Daniel Görges, and Steven Liu. Stability and control of systems with uncertain time-varying sampling period and time delay. In *Proceedings of the 17th IFAC World Congress*, pages 11514–11519, 2008. (Implementation-Aware Control)

[GIL09] Daniel Görges, Michal Izák, and Steven Liu. Optimal control and scheduling of networked control systems. In *Proceedings of the 48th IEEE Conference on Decision and Control and 28th Chinese Control Conference*, pages 5839–5844, 2009. (Optimal Control of Switched Systems, Chapter 3, and Control and Scheduling Codesign, Chapters 7, 8, 10)

[IGL10b] Michal Izák, Daniel Görges, and Steven Liu. Stabilization of systems with variable and uncertain sampling period and time delay. *Nonlinear Analysis: Hybrid Systems*, 4(2):291–305, 2010. (Implementation-Aware Control)

[SGIL10] Stefan Simon, Daniel Görges, Michal Izák, and Steven Liu. Periodic observer design for networked embedded control systems. In *Proceedings of the 2010 American Control Conference*, pages 4253–4258, 2010. (Control and Scheduling Codesign)

[IGL10a] Michal Izák, Daniel Görges, and Steven Liu. Optimal control of networked control systems with uncertain time-varying transmission delay. In *Proceedings of the 2nd IFAC Workshop on Distributed Estimation and Control in Networked Systems*, pages 13–18, 2010. (Implementation-Aware Control)

[GIL11] Daniel Görges, Michal Izák, and Steven Liu. Optimal control and scheduling of switched systems. *IEEE Transactions on Automatic Control*, 56(1):135–140, 2011. (Optimal Control of Switched Systems, Chapter 3)

[AGL11] Sanad Al-Areqi, Daniel Görges, and Steven Liu. Robust control and scheduling codesign for networked embedded control systems. In *Proceedings of the 50th IEEE Conference on Decision and Control and European Control Conference 2011*, pages 3154–3159, 2011. (Control and Scheduling Codesign)

Acknowledgments

This thesis presents the results of my research at the Institute of Control Systems in the Department of Electrical and Computer Engineering at the University of Kaiserslautern.

Foremost, I would like to express my utmost gratitude to Prof. Dr.-Ing. Steven Liu, the head of the Institute of Control Systems, for the excellent supervision of my thesis. He has sparked my passion for control theory, has provided important hints and advice and has always been available for scientific discussions. He has also encouraged the publication of my results and supported trips to conferences and workshops to present my results, to view the most recent results of other researchers and to get in contact with the scientific community. Moreover, I sincerely appreciate the high scientific freedom. All this gave me outstanding opportunities to learn and grow as a researcher.

Furthermore, I would like to thank Prof. Dr. Karl Henrik Johansson for his interest in my thesis, the inspiring discussions and for joining the thesis committee as a reviewer. Thanks also go to Prof. Dr.-Ing. Dr. rer. nat. habil. Wolfgang Kunz for joining the thesis committee as the chair.

My time at the Institute of Control Systems has been very enjoyable and rewarding. For this I am particularly indebted to my colleagues and former colleagues. They created an open atmosphere, leaving much room for both scientific and personal discussions. My sincere gratitude goes to M. Sc. Michal Izák, whom I very much appreciate as a researcher as well as a friend. Our countless fruitful discussions have considerably contributed to this thesis. Special thanks also go to M. Sc. Sanad Al-Areqi, Dr.-Ing. Liang Chen, Dr.-Ing. Jens Kroneis, Dipl.-Ing. Peter Müller, Dr.-Ing. Philipp Münch, Dipl.-Ing. Tim Nagel, Dr.-Ing. Martin Pieschel, Dipl.-Wirtsch.-Ing. Sven Reimann, Dipl.-Ing. Stefan Simon, Dipl.-Ing. Nadine Stegmann-Drüppel, Priv. Doz. Dr.-Ing. habil. Christian Tuttas, M. Sc. Yun Wan, M. Sc. Jianfei Wang and M. Sc. Wei Wu for a very good collaboration and many insightful scientific and personal discussions. Thanks also go to the technicians Swen Becker and Thomas Janz and to the secretary Jutta Lenhardt for providing a very good technical and administrative environment.

Finally, I would like to express my deep gratitude to my parents and family for their encouragement, confidence and patience over all the years. This thesis is dedicated to them.

Kaiserslautern, July 2011 *Daniel Görges*

Notation

Throughout the thesis, scalars are denoted by lower- and upper-case non-bold letters ($a, b, \ldots, A, B, \ldots$), vectors by lower-case bold letters ($\boldsymbol{a}, \boldsymbol{b}, \ldots$), matrices by upper-case bold letters ($\boldsymbol{A}, \boldsymbol{B}, \ldots$) and sets by upper-case double-struck letters ($\mathbb{A}, \mathbb{B}, \ldots$).

Sets

$\mathbb{N}$	Set of positive integers
$\mathbb{N}_0$	Set of non-negative integers
$\mathbb{R}$	Set of real numbers
$\mathbb{R}_0^+$	Set of non-negative real numbers
$\mathbb{R}^-$	Set of negative real numbers
$\mathbb{C}$	Set of complex numbers

Operators

$\boldsymbol{A}^{-1}$	Inverse of matrix $\boldsymbol{A}$
$\boldsymbol{A}^T$	Transpose of matrix $\boldsymbol{A}$
$\boldsymbol{A} \succ 0$	Matrix $\boldsymbol{A} \in \mathbb{R}^{\mathrm{n}\times\mathrm{n}}$ positive definite, i.e. $\boldsymbol{x}^T\boldsymbol{A}\boldsymbol{x} > 0 \; \forall \boldsymbol{x} \in \mathbb{R}^{\mathrm{n}} \setminus \{0\}$
$\boldsymbol{A} \succeq 0$	Matrix $\boldsymbol{A} \in \mathbb{R}^{\mathrm{n}\times\mathrm{n}}$ positive semidefinite, i.e. $\boldsymbol{x}^T\boldsymbol{A}\boldsymbol{x} \geq 0 \; \forall \boldsymbol{x} \in \mathbb{R}^{\mathrm{n}}$
$\boldsymbol{A} \prec 0$	Matrix $\boldsymbol{A} \in \mathbb{R}^{\mathrm{n}\times\mathrm{n}}$ negative definite, i.e. $\boldsymbol{x}^T\boldsymbol{A}\boldsymbol{x} < 0 \; \forall \boldsymbol{x} \in \mathbb{R}^{\mathrm{n}} \setminus \{0\}$
$\boldsymbol{A} \preceq 0$	Matrix $\boldsymbol{A} \in \mathbb{R}^{\mathrm{n}\times\mathrm{n}}$ negative semidefinite, i.e. $\boldsymbol{x}^T\boldsymbol{A}\boldsymbol{x} \leq 0 \; \forall \boldsymbol{x} \in \mathbb{R}^{\mathrm{n}}$
$\mathrm{tr}(\boldsymbol{A})$	Trace of matrix $\boldsymbol{A}$
$\det(\boldsymbol{A})$	Determinant of matrix $\boldsymbol{A}$
$\lambda_{\mathrm{min}}(\boldsymbol{A})$	Minimum eigenvalue of matrix $\boldsymbol{A}$
$\lambda_{\mathrm{max}}(\boldsymbol{A})$	Maximum eigenvalue of matrix $\boldsymbol{A}$
$\mathrm{diag}(\boldsymbol{A}_1, \ldots)$	Block-diagonal matrix with blocks $\boldsymbol{A}_1, \ldots$
$\lVert\boldsymbol{x}\rVert$	Arbitrary p-norm of vector $\boldsymbol{x} \in \mathbb{R}^{\mathrm{n}}$
$\lVert\boldsymbol{x}\rVert_2$	Euclidean norm of vector $\boldsymbol{x} \in \mathbb{R}^{\mathrm{n}}$, i.e. $\lVert\boldsymbol{x}\rVert_2 = \sqrt{\boldsymbol{x}^T\boldsymbol{x}} = \sqrt{x_1^2 + \ldots + x_{\mathrm{n}}^2}$
$\mathrm{E}(X)$	Expected value of random variable X
$\lfloor x \rfloor$	Floor, i.e. $\lfloor x \rfloor$ is the largest integer smaller than or equal to $x \in \mathbb{R}$
$\lceil x \rceil$	Ceil, i.e. $\lceil x \rceil$ is the smallest integer larger than or equal to $x \in \mathbb{R}$
$x \bmod y$	Modulo, i.e. $x \bmod y = x - y \left\lfloor \frac{x}{y} \right\rfloor$ with $x, y \in \mathbb{R}$

$\lvert\mathbb{M}\rvert$	Cardinality of the set $\mathbb{M}$

Others

$\boldsymbol{I}$	Identity matrix
0	Zero matrix
$\begin{pmatrix} \boldsymbol{A} & * \\ \boldsymbol{B} & \boldsymbol{C} \end{pmatrix}$	Symmetric matrix $\begin{pmatrix} \boldsymbol{A} & \boldsymbol{B}^T \\ \boldsymbol{B} & \boldsymbol{C} \end{pmatrix}$
x^*	Variable x determined by optimization

Acronyms

ARE	Algebraic Riccati Equation
BMI	Bilinear Matrix Inequality
CQLF	Common Quadratic Lyapunov Function
DRE	Difference Riccati Equation
DPRE	Discrete-Time Periodic Riccati Equation
ECS	Embedded Control System
FHCS	Finite-Horizon Control and Scheduling (Chapter 2)
$\mathrm{HPCS_{on}}$	Heuristic Periodic Control and Online Scheduling (Section 4.2.4)
iff	if and only if
IHCS	Infinite-Horizon Control and Scheduling
LMI	Linear Matrix Inequality
N'-step RHCS	N'-step RHCS Receding-Horizon Control and Scheduling (Chapter 9)
LQR	Linear-Quadratic Regulator
LTI	Linear Time-Invariant
MAC	Medium Access Control
MIP	Mixed Integer Programming
MLD	Mixed Logical Dynamical
MPC	Model Predictive Control
NCS	Networked Control System
NECS	Networked Embedded Control System
OPP	Optimal Pointer Placement (Section 4.2.4)
$\mathrm{PCS_{off}}$	Periodic Control and Offline Scheduling (Section 4.1)
$\mathrm{PCS_{on}}$	Periodic Control and Online Scheduling (Section 4.2)
PLQR	Periodic Linear-Quadratic Regulator (Section 4.1.2)
PLTV	Periodic Linear Time-Varying
PWA	Piecewise Affine
PWL	Piecewise Linear
PWQ	Piecewise Quadratic
RHC	Receding-Horizon Control

RHCS	Receding-Horizon Control and Scheduling (Chapter 3)
RPCS_{on}	Relaxed Periodic Control and Online Scheduling (Section 4.2.4)
RRHCS	Relaxed Receding-Horizon Control and Scheduling (Chapter 3)
SPPS	Symmetric Periodic Positive Semidefinite
SQLF	Switched Quadratic Lyapunov Function
ZOH	Zero-Order Hold

Linear Matrix Inequalities

Linear matrix inequalities (LMIs) are utilized throughout the thesis. Introductions to LMIs are given in [BEFB94] and [SP05] where also related topics like the S-procedure [BEFB94, Section 2.6.3], [SP05, Section 12.3.4], the Schur complement [BEFB94, pages 7–8], [SP05, Section 12.3.3] and the congruence transformation [SP05, Section 12.3.2] are addressed.

Implementation

Computations and simulations have been performed on a PC with Intel® Core™2 Duo (E8400) processor, 4 GB RAM and Microsoft® Windows Vista® 6.0 (Build 6002, Service Pack 2, 32 bit) operating system using MATLAB® 7.7.0.471 (R2008b,win-32) with the Control System Toolbox™8.2 (R2008b) for solving AREs and YALMIP 3.0 (R20100702) [Löf04] with SeDuMi 1.3 [Stu99] for solving general LMI problems and with SDPT3 4.0 [TTT99, TTT03] for solving the LMI optimization problem (3.28).

Part I

Optimal Control of Switched Systems

1 Introduction

1.1 Hybrid Systems

Many systems involve interactions between continuous dynamics governed by differential or difference equations and discrete dynamics governed by logics. Such systems are called hybrid systems [Wit66]. Hybrid systems arise in various applications, e.g. in networked embedded control systems (Part II), communication networks [BBE+05], automotive systems [BBD+00], power systems [GLM03, BGM05], power electronics [MAB+10] and process control [LTEP96, EKSS00]. Typical examples include cars with gear shifting, mechanical systems with backlash, inverters with diodes and switches, chemical plants with on-off valves and networked embedded control systems with digital components interacting with the analog environment.

Understanding the complex interactions between continuous and discrete dynamics is essential to guarantee stability and performance of hybrid systems. Specific methods for modeling, analysis and control of hybrid systems are strongly required. Since hybrid systems are inherently interdisciplinary, concepts from diverse areas like mathematics, control theory, computer science and communications have to be combined to this end.

Various surveys [AK03, BBE+05, GST09], special issues [AN98, MPSS99, Ant00] and monographs [vdSS99, LL09] on hybrid systems have been published in recent years, underlining the large interest both in academia and industry.

1.2 Switched Systems

Switched systems constitute a very important class [SG05a] and abstraction [Lib03] of hybrid systems. They consist of a set of subsystems and a switching law that defines switching between these subsystems. Formally, a continuous-time switched system is represented by a differential equation

$$\dot{\boldsymbol{x}}(t) = \boldsymbol{f}_{j(k)}(\boldsymbol{x}(t), \boldsymbol{u}(t)), \qquad \boldsymbol{x}(0) = \boldsymbol{x}_0 \tag{1.1}$$

with state vector $\boldsymbol{x}(t) \in \mathbb{R}^{\mathrm{n}}$, control vector $\boldsymbol{u}(t) \in \mathbb{R}^{\mathrm{m}}$ and continuous time $t \in \mathbb{R}_0^+$. Analogously, a discrete-time switched system is represented by a difference equation

$$\boldsymbol{x}(k+1) = \boldsymbol{f}_{j(k)}(\boldsymbol{x}(k), \boldsymbol{u}(k)), \qquad \boldsymbol{x}(0) = \boldsymbol{x}_0 \tag{1.2}$$

with state vector $\boldsymbol{x}(k) \in \mathbb{R}^{\mathrm{n}}$, control vector $\boldsymbol{u}(k) \in \mathbb{R}^{\mathrm{m}}$ and discrete time $k \in \mathbb{N}_0$. The generally nonlinear functions $\boldsymbol{f}_{j(k)} : \mathbb{R}^{\mathrm{n}} \times \mathbb{R}^{\mathrm{m}} \to \mathbb{R}^{\mathrm{n}}$ which are indexed by the *switching index* $j(k) \in \mathbb{M} = \{1, \ldots, M\}$ are denoted as *subsystems*. Switching between the subsystems at time instants t_k, $k \in \mathbb{N}_0$ is defined by the *switching sequence* $j : \mathbb{N}_0 \to \mathbb{M}$.

Switched systems are generally classified by their switching behavior:

- *Autonomous Switching*

  Autonomous switching is forced internally and can be *state- and input-dependent*, i.e. $j(k) = f(\boldsymbol{x}(t), \boldsymbol{u}(t))$ (continuous time) or $j(k) = f(\boldsymbol{x}(k), \boldsymbol{u}(k))$ (discrete time), or *time-dependent*, i.e. $j(k) = f(t)$ (continuous time) or $j(k) = f(k)$ (discrete time). State- and input-dependent switching is encountered e.g. in inverters with diodes, time-dependent switching e.g. in inverters using pulse-width modulation (PWM).

- *Controlled Switching*

  Controlled switching, also denoted as *scheduling*, is forced externally and can be *arbitrary* or *constrained*. Controlled switching is encountered e.g. in inverters with switches. Various constraints on switching can be considered. Examples include *dwell times* $\tau_d \in \mathbb{R}_0^+$ (continuous time) or $k_d \in \mathbb{N}_0$ (discrete time) specifying the minimum time between switches, i.e. $t_{k+1} - t_k \geq \tau_d$ or $j(k) = j(k + k')$ for all $k \leq k' \leq k_d$, or *periodicity* constraints, i.e. $t_{k+p} - t_k = \text{const.}$ and $j(k) = j(k + p)$ for $p \in \mathbb{N}_0$ and all $k \in \mathbb{N}_0$. Constraints on switching can also be dictated by the application. For example the gears in a car can not be shifted arbitrarily.

This classification is not always unique. Specifically, combinations of autonomous and controlled switching as well as state- and input-dependent, time-dependent, arbitrary and constrained switching can occur. For example inverters with both diodes and switches may involve autonomous state- and input-dependent switching (diodes), autonomous time-dependent switching (switches controlled internally via PWM) as well as controlled arbitrary switching (switches controlled externally). Moreover, state- and input-dependent switching is partly forced externally via $\boldsymbol{u}(t)$ (continuous time) or $\boldsymbol{u}(k)$ (discrete time) in an implicit way.

Continuous-time switched systems can furthermore exhibit *Zeno behavior* [ZJLS01], i.e. infinitely many switches in a finite time, and *impulse effects*, i.e. $\boldsymbol{x}(t_k^-) \neq \boldsymbol{x}(t_k^+)$. These effects do not arise in discrete-time switched systems.

An important class of switched systems are discrete-time switched linear systems with controlled arbitrary switching. A discrete-time switched linear system is represented by the difference equation

$$\boldsymbol{x}(k+1) = \boldsymbol{A}_{j(k)}\boldsymbol{x}(k) + \boldsymbol{B}_{j(k)}\boldsymbol{u}(k), \quad \boldsymbol{x}(0) = \boldsymbol{x}_0 \tag{1.3}$$

where the pair $(\boldsymbol{A}_{j(k)}, \boldsymbol{B}_{j(k)})$ comprising the system matrix $\boldsymbol{A}_{j(k)} \in \mathbb{R}^{\mathrm{n}\times\mathrm{n}}$ and the input matrix $\boldsymbol{B}_{j(k)} \in \mathbb{R}^{\mathrm{n}\times\mathrm{m}}$ is denoted as subsystem.

Another important class of switched systems are discrete-time piecewise affine (PWA) systems with autonomous state- and input-dependent switching. A discrete-time PWA system is represented by the difference equation

$$\boldsymbol{x}(k+1) = \boldsymbol{A}_{j(k)}\boldsymbol{x}(k) + \boldsymbol{B}_{j(k)}\boldsymbol{u}(k) + \boldsymbol{g}_{j(k)} \quad \text{for} \quad \begin{pmatrix}\boldsymbol{x}(k)\\ \boldsymbol{u}(k)\end{pmatrix} \in \mathcal{X}_{j(k)}, \quad \boldsymbol{x}(0) = \boldsymbol{x}_0 \tag{1.4}$$

where the pair $(\boldsymbol{A}_{j(k)}, \boldsymbol{B}_{j(k)}, \boldsymbol{g}_{j(k)})$ comprising the system matrix $\boldsymbol{A}_{j(k)} \in \mathbb{R}^{\mathrm{n}\times\mathrm{n}}$, the input matrix $\boldsymbol{B}_{j(k)} \in \mathbb{R}^{\mathrm{n}\times\mathrm{m}}$ and the affine term $\boldsymbol{g}_{j(k)} \in \mathbb{R}^{\mathrm{n}}$ is denoted as subsystem. The switching sequence $j(k)$, $k \in \mathbb{N}_0$ is defined implicitly by the regions $\mathcal{X}_{j(k)} \subset \mathbb{R}^{\mathrm{n}+\mathrm{m}}$ of the state and input space. Usually polyhedral regions $\mathcal{X}_{j(k)}$ are considered. PWA systems are under mild conditions equivalent to various classes of hybrid systems such as mixed logical dynamical (MLD) systems [BM99] as shown in [HDB01] and are therefore widely studied. The PWA system (1.4) reduces to a piecewise linear (PWL) system if $\boldsymbol{g}_{j(k)} = \boldsymbol{0}$ for all $j(k) \in \mathbb{M}$.

This thesis is focused on discrete-time switched linear systems (1.3) with controlled and arbitrary switching. For brevity the discrete-time switched linear system (1.3) is commonly denoted as switched system in the following.

1.3 Stability and Stabilization of Switched Systems

The stability and stabilization of switched systems has been intensely studied in recent years. Excellent surveys are given in [LM99, DBPL00, Lib03, BGLM05, SG05a, SG05b, SWM+07, LA09]. Therefore, only basic concepts for stability analysis and stabilization of discrete-time switched linear systems are reviewed in the following sections. The concepts can be classified by stability under arbitrary switching (Section 1.3.2), stability under constrained switching including state-dependent switching (Section 1.3.3) and stabilization (Section 1.3.4).

Discrete-time switched linear systems are inherently non-autonomous systems. Thus, Lyapunov stability theory for discrete-time non-autonomous systems is reviewed first in the following section.

1.3.1 Stability of Non-Autonomous Systems

Consider the discrete-time non-autonomous system

$$\boldsymbol{x}(k+1) = \boldsymbol{f}(\boldsymbol{x}(k), k), \quad \boldsymbol{x}(k_0) = \boldsymbol{x}_0 \tag{1.5}$$

where $\boldsymbol{x}(k) \in \mathbb{R}^{\mathrm{n}}$ is the state vector, $k \in \mathbb{N}_0$ is the discrete time, $k_0 \in \mathbb{N}_0$ is the initial time and $\boldsymbol{f} : \mathbb{R}^{\mathrm{n}} \times \mathbb{N}_0 \to \mathbb{R}^{\mathrm{n}}$ is a generally nonlinear function. Assume that $\boldsymbol{x}_e = \boldsymbol{0}$ is an equilibrium point of (1.5), i.e. $\boldsymbol{f}(\boldsymbol{0}, k) = \boldsymbol{0}$ for all $k \in \mathbb{N}_0$. This assumption is not restrictive since a stability analysis for other equilibrium points can always be reformulated to a stability analysis for the equilibrium point $\boldsymbol{x}_e = \mathbf{0}$ by a change of variables. The stability of the equilibrium point $\boldsymbol{x}_e = \mathbf{0}$ in the sense of Lyapunov is characterized by the following

Definition 1.1 *The equilibrium point $\boldsymbol{x}_e = \mathbf{0}$ of the discrete-time non-autonomous system (1.5) is*

- *stable at k_0 if for each $\varepsilon > 0$ there exists a $\delta = \delta(\varepsilon, k_0) > 0$ such that*

$$||\boldsymbol{x}(k_0)|| < \delta \ \Rightarrow \ ||\boldsymbol{x}(k)|| < \varepsilon \quad \forall k \geq k_0, \tag{1.6}$$

- *uniformly stable if for each $\varepsilon > 0$ there exists a $\delta = \delta(\varepsilon) > 0$ independent of k_0 such that (1.6) is fulfilled,*
- *asymptotically stable at k_0 if it is stable and there exists a $\delta'(k_0) > 0$ such that*

$$||\boldsymbol{x}(k_0)|| < \delta' \ \Rightarrow \ \lim_{k \to \infty} ||\boldsymbol{x}(k)|| = \mathbf{0}, \tag{1.7}$$

- *uniformly asymptotically stable if it is stable and there exists a $\delta' > 0$ independent of k_0 such that (1.7) is fulfilled uniformly in k_0, i.e. for each $\varepsilon' > 0$ there exists a $K = K(\varepsilon')$ independent of k_0 such that*

$$||\boldsymbol{x}(k_0)|| < \delta' \ \Rightarrow \ ||\boldsymbol{x}(k)|| < \varepsilon' \quad \forall k \geq k_0 + K, \tag{1.8}$$

- *globally uniformly asymptotically stable if it is uniformly asymptotically stable for all $\boldsymbol{x}(k_0) \in \mathbb{R}^{\mathrm{n}}$,*
- *unstable if it is not stable.*

Remark 1.1. The stability of the equilibrium point $\boldsymbol{x}_e = \mathbf{0}$ generally depends on the initial time k_0 which is usually not desirable. This dependence can be removed by considering uniform stability.

Checking the conditions given in Definition 1.1 for general discrete-time non-autonomous systems (1.5) is difficult and no universal method exists. One important method is Lyapunov's direct method. The cornerstone of Lyapunov's direct method are positive definite, positive semidefinite, negative definite and negative semidefinite functions which are defined as follows:

Definition 1.2 *A function $V : \mathbb{D} \to \mathbb{R}$ is*

- *positive semidefinite in $\mathbb{D} \subset \mathbb{R}^{\mathrm{n}}$ if*

  (1) $V(\mathbf{0}) = 0$

  (2) $V(\boldsymbol{x}(k)) \geq 0 \quad \forall \boldsymbol{x}(k) \in \mathbb{D} \setminus \{\mathbf{0}\}$,

- *positive definite in $\mathbb{D} \subset \mathbb{R}^{\mathrm{n}}$ if (2) is replaced by*

(2') $V(\boldsymbol{x}(k)) > 0 \quad \forall \boldsymbol{x}(k) \in \mathbb{D} \setminus \{\boldsymbol{0}\}$,

- *negative definite (semidefinite) in $\mathbb{D} \subset \mathbb{R}^n$ if $-V$ is positive definite (semidefinite).*

Definition 1.3 *A function $V : \mathbb{D} \times \mathbb{N}_0 \to \mathbb{R}$ is*

- *positive semidefinite in $\mathbb{D} \subset \mathbb{R}^n$ if*

  (1) $V(\boldsymbol{0}, k) = 0 \quad \forall k \in \mathbb{N}_0$

  (2) $V(\boldsymbol{x}(k), k) \geq 0 \quad \forall \boldsymbol{x}(k) \in \mathbb{D} \setminus \{\boldsymbol{0}\} \quad \forall k \in \mathbb{N}_0$,

- *positive definite in $\mathbb{D} \subset \mathbb{R}^n$ if*

  (1) $V(\boldsymbol{0}, k) = 0 \quad \forall k \in \mathbb{N}_0$

  (2) there exists a positive definite function $V_1 : \mathbb{D} \to \mathbb{R}$ independent of k such that

$$V_1(\boldsymbol{x}(k)) \leq V(\boldsymbol{x}(k), k) \quad \forall \boldsymbol{x}(k) \in \mathbb{D} \quad \forall k \in \mathbb{N}_0,$$

- *negative definite (semidefinite) in $\mathbb{D} \subset \mathbb{R}^n$ if $-V$ is positive definite (semidefinite),*

- *decrescent if there exists a positive definite function $V_2 : \mathbb{D} \to \mathbb{R}$ independent of k such that*

$$V(\boldsymbol{x}(k), k) \leq V_2(\boldsymbol{x}(k)) \quad \forall \boldsymbol{x}(k) \in \mathbb{D} \quad \forall k \in \mathbb{N}_0,$$

- *radially unbounded if there exists a positive definite function $V_1 : \mathbb{D} \to \mathbb{R}$ independent of k with $V_1(\boldsymbol{x}(k)) \to \infty$ as $||\boldsymbol{x}(k)|| \to \infty$ such that*

$$V_1(\boldsymbol{x}(k)) \leq V(\boldsymbol{x}(k), k) \quad \forall \boldsymbol{x}(k) \in \mathbb{D} \quad \forall k \in \mathbb{N}_0.$$

Lyapunov's direct method can then be formalized as

Theorem 1.4 *If in a neighborhood $\mathbb{D} \subset \mathbb{R}^n$ of the equilibrium point $\boldsymbol{x}_e = \boldsymbol{0}$ of the discrete-time non-autonomous system (1.5) there exists a function $V : \mathbb{D} \times \mathbb{N}_0 \to \mathbb{R}$ such that*

(1) $V(\boldsymbol{x}(k), k)$ is positive definite

(2) $\Delta V(\boldsymbol{x}(k), k) = V(\boldsymbol{x}(k+1), k+1) - V(\boldsymbol{x}(k), k)$ is negative semidefinite,

then the equilibrium point is stable. If furthermore

(3) $V(\boldsymbol{x}(k), k)$ is decrescent,

then the equilibrium point is uniformly stable. If furthermore

(2') $\Delta V(\boldsymbol{x}(k), k) = V(\boldsymbol{x}(k+1), k+1) - V(\boldsymbol{x}(k), k)$ is negative definite,

then the equilibrium point is uniformly asymptotically stable. If furthermore $\mathbb{D} = \mathbb{R}^n$ and

(4) $V(\boldsymbol{x}(k), k)$ is radially unbounded,

then the equilibrium point is globally uniformly asymptotically stable.

PROOF. The proof can be deduced from the proof for continuous-time non-autonomous systems, see e.g. [SL91, Section 4.2.1], [Vid02, Section 5.3] and [Mar03, Section 4.4]. See also [Vid02, Section 5.9] and [Mar03, Section 4.10] for a discussion on discrete-time non-autonomous systems (1.5). □

Remark 1.2. A function $V(\boldsymbol{x}(k), k)$ fulfilling at least conditions (1) and (2) of Theorem 1.4 is called Lyapunov function for the discrete-time non-autonomous system (1.5).

Remark 1.3. Global uniform asymptotic stability of the equilibrium point $\boldsymbol{x}_e = \boldsymbol{0}$ of the discrete-time non-autonomous system (1.5) implies uniqueness of this equilibrium point. The discrete-time non-autonomous system (1.5) is therefore commonly denoted itself as globally uniformly asymptotically stable.

Remark 1.4. Uniform asymptotic stability can be strengthened to exponential stability, see [Vid02, Section 5.9] for a definition. Note that most stability conditions given in this thesis can also be formulated w.r.t. exponential stability. For conciseness, however, only uniform asymptotic stability is studied.

Remark 1.5. For discrete-time periodic systems (1.5) with $\boldsymbol{f}(\boldsymbol{x}, k) = \boldsymbol{f}(\boldsymbol{x}, k+p)$ for all $\boldsymbol{x} \in \mathbb{R}^{\mathrm{n}}$, all $k \in \mathbb{N}_0$ and the period $p \in \mathbb{N}_0$, the stability and the uniform stability of the equilibrium point $\boldsymbol{x}_e = \boldsymbol{0}$ are equivalent, see e.g. [Vid02, p. 143].

Remark 1.6. The stability in the sense of Lyapunov relates to an unforced system, i.e. $\boldsymbol{u}(k) = \boldsymbol{0}\ \forall k \in \mathbb{N}_0$, or a closed-loop system, i.e. $\boldsymbol{u}(k) = \boldsymbol{f}_{\mathrm{C}}(\boldsymbol{x}(k), k)$ with the function $\boldsymbol{f}_{\mathrm{C}} : \mathbb{R}^{\mathrm{n}} \times \mathbb{N}_0 \rightarrow \mathbb{R}^{\mathrm{m}}$ describing some controller. All stability conditions given in this thesis are formulated in this sense.

1.3.2 Stability under Arbitrary Switching

A necessary condition for stability of the switched system (1.3) under arbitrary switching is that all subsystems $\boldsymbol{A}_{j(k)}$, $j(k) \in \mathbb{M}$ must be individually globally asymptotically stable. Otherwise, an unstable subsystem may be selected at each time instant k under arbitrary switching, rendering the switched system unstable.

A sufficient condition for stability of the switched system (1.3) can be formulated based on a *common quadratic Lyapunov function* (CQLF).

Theorem 1.5 (Stability Condition based on a CQLF) *The switched system (1.3) is globally uniformly asymptotically stable if there exists a symmetric matrix $\boldsymbol{S} \succ \boldsymbol{0}$ such that*

$$\boldsymbol{A}_v^T \boldsymbol{S} \boldsymbol{A}_v - \boldsymbol{S} \prec 0 \quad \forall v \in \mathbb{M}. \tag{1.9}$$

The quadratic function $V(\boldsymbol{x}(k), k) = \boldsymbol{x}^T(k) \boldsymbol{S} \boldsymbol{x}(k)$ is then a CQLF for the switched system (1.3).

PROOF. The function $V(\boldsymbol{x}(k),k) = \boldsymbol{x}^T(k)\boldsymbol{S}\boldsymbol{x}(k)$ is positive definite, decrescent and radially unbounded since

$$\alpha_1||\boldsymbol{x}(k)||_2^2 \leq V(\boldsymbol{x}(k),k) \leq \alpha_2||\boldsymbol{x}(k)||_2^2 \quad \forall \boldsymbol{x}(k) \in \mathbb{R}^n \quad \forall k \in \mathbb{N}_0 \tag{1.10}$$

with

$$\alpha_1 = \lambda_{\min}(\boldsymbol{S}) > 0, \quad \alpha_2 = \lambda_{\max}(\boldsymbol{S}) > 0 \tag{1.11}$$

resulting from Lemma A.2 for $\boldsymbol{S} \succ \boldsymbol{0}$. The difference of the function $V(\boldsymbol{x}(k),k)$ along trajectories of the switched system (1.3) given by

$$\begin{aligned}\Delta V(\boldsymbol{x}(k),k) &= \boldsymbol{x}^T(k+1)\boldsymbol{S}\boldsymbol{x}(k+1) - \boldsymbol{x}^T(k)\boldsymbol{S}\boldsymbol{x}(k) \\ &= \boldsymbol{x}^T(k)\left(\boldsymbol{A}_{j(k)}^T\boldsymbol{S}\boldsymbol{A}_{j(k)} - \boldsymbol{S}\right)\boldsymbol{x}(k)\end{aligned} \tag{1.12}$$

is negative definite if (1.9) is fulfilled since

$$\Delta V(\boldsymbol{x}(k),k) \leq \alpha_3||\boldsymbol{x}(k)||_2^2 \quad \forall \boldsymbol{x}(k) \in \mathbb{R}^n \quad \forall k \in \mathbb{N}_0 \tag{1.13}$$

with

$$\alpha_3 = \max_{j(k)\in\mathbb{M}} \lambda_{\max}\left(\boldsymbol{A}_{j(k)}^T\boldsymbol{S}\boldsymbol{A}_{j(k)} - \boldsymbol{S}\right) < 0, \tag{1.14}$$

resulting again from Lemma A.2. This completes the proof. □

The stability condition in Theorem 1.5 corresponds to an LMI feasibility problem which can be efficiently solved numerically, see e.g. [BV04]. Besides numerical methods also algebraic conditions for the existence of a CQLF have been proposed, e.g. based on Lie algebra [Lib03, Section 2.2] or matrix pencils [SWM+07, Section 4]. These conditions are, however, usually only applicable for switched systems with special structure and partly difficult to evaluate as discussed in [SWM+07]. Such conditions have furthermore only been studied for continuous-time switched linear systems. Considering common Lyapunov functions has been legitimated by several *converse Lyapunov theorems*. These suggest, however, that common non-quadratic Lyapunov functions should be studied. E.g. in [MP89, Bla95] it has been shown that if the switched system (1.3) is globally uniformly asymptotically stable under arbitrary switching, then there exists a strictly convex, homogeneous of degree 2 and quasi-quadratic Lyapunov function or a polyhedral Lyapunov function. Noteworthy, these results have been formulated for systems with uncertainties but also apply to switched systems as outlined in [LA09, Section II-D]. Examining such Lyapunov functions in detail is a promising direction for future research.

A sufficient condition for stability of the switched system (1.3) can also be formulated based on a *switched quadratic Lyapunov function* (SQLF).

Theorem 1.6 (Stability Condition based on an SQLF) *The switched system (1.3) is globally uniformly asymptotically stable if there exist symmetric matrices $\boldsymbol{S}_v \succ \boldsymbol{0}$, $v \in \mathbb{M}$ such that*

$$\boldsymbol{A}_v^T\boldsymbol{S}_w\boldsymbol{A}_v - \boldsymbol{S}_v \prec \boldsymbol{0} \quad \forall (v,w) \in \mathbb{M}^2 \tag{1.15}$$

where $(v, w) \in \mathbb{M}^2$ *is a pair of consecutive switching indices, i.e.* $v = j(k)$ *and* $w = j(k+1)$. *The quadratic function* $V(\boldsymbol{x}(k), k) = \boldsymbol{x}^T(k)\boldsymbol{S}_{j(k)}\boldsymbol{x}(k)$ *is then an SQLF for the switched system (1.3).*

PROOF. See the proof of Theorem 2 in [DRI02]. □

The stability condition in Theorem 1.6 again corresponds to an LMI feasibility problem and reduces to the stability condition in Theorem 1.5 if $\boldsymbol{S}_v = \boldsymbol{S}_w$ for all $(v, w) \in \mathbb{M}^2$. The stability condition based on an SQLF is therefore less conservative than the stability condition based on a CQLF.

Necessary and sufficient conditions for stability of the switched system (1.3) under arbitrary switching are still under research. Some important results have been reported in [BPD93, LA05] and [SWM⁺07, Section 4.7].

1.3.3 Stability under Constrained Switching

The stability of the switched system (1.3) under constrained switching can be examined based on *multiple Lyapunov functions* [Bra98, DBPL00]. The idea consists in assigning a Lyapunov function $V_{j(k)}(\boldsymbol{x}(k))$ to each subsystem or alternatively each state-space region. These Lyapunov functions are then concatenated to a non-traditional Lyapunov function. The Lyapunov functions $V_{j(k)}(\boldsymbol{x}(k))$ may not always decrease. A decrease is only required at consecutive time instants t_k^+ and $t_{k+k'}^+$ (or t_k^- and $t_{k+k'}^-$) for which $j(k) = j(k + k')$ holds. This is illustrated in Figure 1.1.

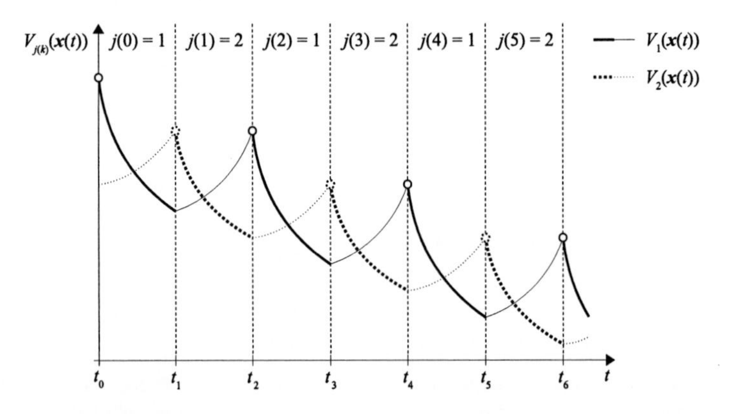

Figure 1.1: Multiple Lyapunov functions for a continuous-time switched system ($M = 2$). The decrease at consecutive time instants t_k^+ and t_{k+2}^+ is highlighted by circles.

Methods for constructing multiple Lyapunov functions for PWA systems have been proposed in [JR98, MFM00, FCMM02]. Piecewise quadratic (PWQ) Lyapunov functions have been introduced to this end, constituting an important class of multiple Lyapunov functions.

Theorem 1.7 (Stability Condition based on a PWQ Lyapunov Function) *The PWA system (1.4) with $\boldsymbol{g}_v = \boldsymbol{0}$ for all $v \in \mathbb{M}$ and finite $|\mathbb{M}|$ is globally uniformly asymptotically stable if there exists a PWQ Lyapunov function*

$$V(\boldsymbol{x}(k)) = \boldsymbol{x}^T(k)\boldsymbol{S}_v\boldsymbol{x}(k) > 0 \quad \forall \boldsymbol{x}(k) \in \mathcal{X}_v \setminus \{\boldsymbol{0}\},\ v \in \mathbb{M} \tag{1.16}$$

whose difference along trajectories of the PWA system is negative definite, i.e.

$$\begin{aligned} \Delta V(\boldsymbol{x}(k)) &= V(\boldsymbol{x}(k+1)) - V(\boldsymbol{x}(k)) \\ &= \boldsymbol{x}^T(k+1)\boldsymbol{S}_w\boldsymbol{x}(k+1) - \boldsymbol{x}^T(k)\boldsymbol{S}_v\boldsymbol{x}(k) \\ &= \boldsymbol{x}^T(k)\left(\boldsymbol{A}_v^T\boldsymbol{S}_w\boldsymbol{A}_v - \boldsymbol{S}_v\right)\boldsymbol{x}(k) \\ &< 0 \end{aligned} \tag{1.17}$$

for all $\boldsymbol{x}(k) \in \mathcal{X}_v \setminus \{\boldsymbol{0}\}$ and all $\boldsymbol{x}(k+1) \in \mathcal{X}_w$ where $(v, w) \in \mathbb{M}^2$ is a pair of consecutive switching indices, i.e. $v = j(k)$ and $w = j(k+1)$.

PROOF. See the proof of Theorem 1 in [FCMM02] and the discussion in [MFM00]. □

Remark 1.7. An extension to PWA systems (1.4) with $\boldsymbol{g}_v \neq \boldsymbol{0}$ for some (but not all) $v \in \mathbb{M}$ is straightforward, see [FCMM02].

Finding a PWQ Lyapunov function is not trivial. Usually one must resort to more conservative conditions. Methods for constructing PWQ Lyapunov functions for PWA systems with polyhedral regions $\mathcal{X}_v$ have been proposed in [JR98, MFM00, FCMM02]. These methods are based on CQLF or SQLF with relaxations to regard the regions. The resulting stability conditions are formulated as LMI feasibility problems utilizing the S-procedure.

Besides PWQ Lyapunov functions also piecewise polynomial Lyapunov functions and PWA Lyapunov functions have been studied for PWA systems, see [BGLM05] for a survey. Moreover, methods for constructing multiple Lyapunov functions for switched systems with dwell time have been investigated, see e.g. [HM99].

1.3.4 Stabilization

The stabilization of the switched system (1.3) can be studied from various perspectives.

First, designing a stabilizing control sequence $\boldsymbol{u}(k)$, $k \in \mathbb{N}_0$ for an arbitrary or constrained switching sequence $j(k)$, $k \in \mathbb{N}_0$ can be considered. A generic approach to design a stabilizing switched state feedback control law for an arbitrary switching sequence based on an SQLF is given by

Theorem 1.8 (Stabilization based on an SQLF) *Consider the switched state feedback control law*

$$\boldsymbol{u}(k) = -\boldsymbol{K}_{j(k)}\boldsymbol{x}(k) \tag{1.18}$$

with the switched feedback matrices $\boldsymbol{K}_{j(k)}$, $j(k) \in \mathbb{M}$. *The switched system (1.3) under the switched state feedback control law (1.18), precisely the closed-loop system*

$$\boldsymbol{x}(k+1) = (\boldsymbol{A}_{j(k)} - \boldsymbol{B}_{j(k)}\boldsymbol{K}_{j(k)})\boldsymbol{x}(k), \tag{1.19}$$

is globally uniformly asymptotically stable for an arbitrary switching sequence $j(k), k \in \mathbb{N}_0$ *if there exist symmetric matrices* $\boldsymbol{Z}_v \succ \boldsymbol{0}$ *and matrices* $\boldsymbol{W}_v$ *such that*

$$\begin{pmatrix} \boldsymbol{Z}_v & * \\ \boldsymbol{A}_v\boldsymbol{Z}_v - \boldsymbol{B}_v\boldsymbol{W}_v & \boldsymbol{Z}_w \end{pmatrix} \succ \boldsymbol{0} \quad \forall (v,w) \in \mathbb{M}^2 \tag{1.20}$$

where $(v,w) \in \mathbb{M}^2$ *is a pair of consecutive switching indices, i.e.* $v{=}j(k)$ *and* $w{=}j(k{+}1)$. *The stabilizing switched feedback matrices are given by* $\boldsymbol{K}_v = \boldsymbol{W}_v\boldsymbol{Z}_v^{-1}$.

PROOF. Assume that (1.20) is fulfilled. The Schur complement of (1.20) is given by

$$\boldsymbol{Z}_v - (\boldsymbol{A}_v\boldsymbol{Z}_v - \boldsymbol{B}_v\boldsymbol{W}_v)^T\boldsymbol{Z}_w^{-1}(\boldsymbol{A}_v\boldsymbol{Z}_v - \boldsymbol{B}_v\boldsymbol{W}_v) \succ 0. \tag{1.21}$$

Substituting $\boldsymbol{W}_v = \boldsymbol{K}_v\boldsymbol{Z}_v$ and $\boldsymbol{Z}_v = \boldsymbol{S}_v^{-1}$ into (1.21) leads to

$$\boldsymbol{S}_v^{-1} - (\boldsymbol{A}_v\boldsymbol{S}_v^{-1} - \boldsymbol{B}_v\boldsymbol{K}_v\boldsymbol{S}_v^{-1})^T\boldsymbol{S}_w(\boldsymbol{A}_v\boldsymbol{S}_v^{-1} - \boldsymbol{B}_v\boldsymbol{K}_v\boldsymbol{S}_v^{-1}) \succ 0. \tag{1.22}$$

Pre- and post-multiplying (1.22) by $\boldsymbol{S}_v$ and reordering yields

$$(\boldsymbol{A}_v - \boldsymbol{B}_v\boldsymbol{K}_v)^T\boldsymbol{S}_w(\boldsymbol{A}_v - \boldsymbol{B}_v\boldsymbol{K}_v) - \boldsymbol{S}_v \prec 0. \tag{1.23}$$

Global uniform asymptotic stability of the closed-loop system (1.19) now immediately follows from Theorem 1.6. Specifically, the function $V(\boldsymbol{x}(k),k) = \boldsymbol{x}^T(k)\boldsymbol{S}_{j(k)}\boldsymbol{x}(k)$ is an SQLF for the closed-loop system (1.19). □

Remark 1.8. A necessary condition for the stabilizability of the switched system (1.3) by a control sequence $\boldsymbol{u}(k)$, $k \in \mathbb{N}_0$ under an arbitrary switching sequence $j(k)$, $k \in \mathbb{N}_0$ and hence for the existence of a switched state feedback control law (1.18) guaranteeing global uniform asymptotic stability of the closed-loop system (1.19) is that all subsystems $(\boldsymbol{A}_{j(k)},\boldsymbol{B}_{j(k)})$, $j(k) \in \mathbb{M}$ are individually stabilizable. Otherwise, a non-stabilizable subsystem may be selected at each time instant k under an arbitrary switching sequence, rendering the switched system non-stabilizable. Note that this condition is often not necessary for stabilizability of the switched system (1.3) under a constrained switching sequence. This, however, strongly depends on the constraints on the switching sequence.

This approach can be extended in various directions, e.g. to static output feedback [DRI02] and for switched systems with polytopic and norm-bounded uncertainty [IGL08, IGL10a, IGL10b]. Similar methods have furthermore been proposed for stabilization [MFM00] and $\mathcal{H}_\infty$ control [CM02] of PWA systems (1.4).

Second, designing a stabilizing switching sequence $j(k)$, $k \in \mathbb{N}_0$ for the unforced switched system (1.3), i.e. $\boldsymbol{u}(k) = \boldsymbol{0}\ \forall k \in \mathbb{N}_0$, can be considered. In [GC06] the state feedback switching law

$$j(k) = \arg \min_{j(k)\in\mathbb{M}} \boldsymbol{x}^T(k)\boldsymbol{P}_{j(k)}\boldsymbol{x}(k) \tag{1.24}$$

has been proposed and a bilinear matrix inequality (BMI) problem has been formulated to determine symmetric matrices $\boldsymbol{P}_{j(k)} \succ \boldsymbol{0}$ such that the unforced switched system (1.3) is globally uniformly asymptotically stable under the state feedback switching law (1.24). Similar switching laws have been studied in [Pet03] for continuous-time switched linear systems. The concepts presented in Chapters 2 to 4 of this thesis lead to state feedback switching laws similar to (1.24). However, not matrices $\boldsymbol{P}_{j(k)}$ indexed by the switching index $j(k) \in \mathbb{M}$, but matrices $\boldsymbol{P}_l$ or $\boldsymbol{P}_j(0)$ indexed by separate indices $l \in \mathbb{L}_0$ and $j \in \mathbb{J}_p$ are considered. These matrices are determined by optimization, providing a new systematic and efficient way to parametrize the state feedback switching law (1.24).

Third, designing both a stabilizing control sequence $\boldsymbol{u}(k)$, $k \in \mathbb{N}_0$ and a stabilizing switching sequence $j(k)$, $k \in \mathbb{N}_0$ can be considered. This problem has been primarily investigated in terms of optimal control and scheduling of switched systems which is discussed in the following section.

1.4 Optimal Control of Switched Systems

1.4.1 Problem Formulation

Optimal control and scheduling of switched systems (1.3) is concerned with jointly designing a control sequence $\boldsymbol{u}^*(k)$ and a switching sequence $j^*(k)$ such that a cost function is minimized. Within this thesis the discrete-time quadratic cost function

$$J_N = \boldsymbol{x}^T(N)\boldsymbol{Q}_0\boldsymbol{x}(N) + \sum_{k=0}^{N-1} \left[\boldsymbol{x}^T(k)\boldsymbol{Q}_{1j(k)}\boldsymbol{x}(k) + \boldsymbol{u}^T(k)\boldsymbol{Q}_{2j(k)}\boldsymbol{u}(k)\right] \tag{1.25}$$

with the symmetric terminal weighting matrix $\boldsymbol{Q}_0 \succeq \boldsymbol{0}$, the symmetric switched weighting matrices $\boldsymbol{Q}_{1j(k)} \succeq \boldsymbol{0}$ and $\boldsymbol{Q}_{2j(k)} \succ \boldsymbol{0}$ with $j(k) \in \mathbb{M}$ and the time horizon N is considered. Quadratic cost functions have been proposed by Kalman [Kal60] and since then have become very popular due to permitting an accessible interpretation, an easy tuning and often also an analytical solution of the related optimization problem.

The optimization problem can be formalized as

Problem 1.9 *For the switched system (1.3) find a control sequence $\boldsymbol{u}^*(k)$ and a switching sequence $j^*(k)$ with $k = 0, \ldots, N-1$ such that the cost function (1.25) is minimized, i.e.*

$$\min_{\substack{\boldsymbol{u}(0),\ldots,\boldsymbol{u}(N-1)\\ j(0),\ldots,j(N-1)}} J_N \qquad \text{subject to (1.3).} \tag{1.26}$$

Methods for solving Problem 1.9 can be categorized into

- Finite-Horizon Control and Scheduling ($N < \infty$)
- Infinite-Horizon Control and Scheduling ($N = \infty$)
- Receding-Horizon Control and Scheduling ($N < \infty$, receding horizon principle)
- Periodic Control and Scheduling ($N = \infty$, $j(k)$ periodic).

In the following sections general concepts are presented, related work is reviewed and contributions of this thesis are outlined following this categorization.

1.4.2 Finite-Horizon Control and Scheduling

The solution of Problem 1.9 for a finite time horizon N can be deduced from linear-quadratic control theory for discrete-time linear time-varying systems which has been introduced by Kalman in his seminal work [Kal60] (see [KS72, Chapter 6] for the discrete-time formulation): The switching sequence $j(k)$ is fixed; then the optimal control sequence $\boldsymbol{u}^*(k)$ is determined from a difference Riccati equation (DRE). This is repeated for all possible switching sequences $j(k) \in \mathbb{M}^N$, i.e. an explicit enumeration is done. The optimal switching sequence $j^*(k)$ is determined by searching the switching sequence which leads to the minimum cost J_N^* among all possible switching sequences $j(k) \in \mathbb{M}^N$. This procedure is formalized in Section 2.2 utilizing dynamic programming [Bel57, BD62, Ber05a, ÅW90]. The number of possible switching sequences is given by $|\mathbb{M}|^N$ and therefore grows exponentially with the time horizon N. The solution of Problem 1.9 via explicit enumeration thus becomes computationally intractable very rapidly. An exponential complexity generally arises in optimal control of hybrid systems and poses a major challenge. Several methods to reduce the computational complexity have been proposed: In [LB02, ZH08], dynamic programming with pruning has been considered. The idea consists in removing partial solutions during dynamic programming which do not contribute to the optimal solution. Similar methods have been studied in [BBBM05] for optimal control of PWA systems. In [LR02, LR06, Ran06], relaxed dynamic programming has been introduced. The approach is conceptually equivalent to dynamic programming with pruning; optimality is, however, relaxed within predefined bounds. Similar results have been reported in [ZAH09]. Formulations for switched systems with stochastic disturbances are given in [LR01, LB02, ZHL10]. For optimal control of MLD and PWA systems, mixed integer programming (MIP) has been studied in [BM99, Bor03]. MIP is well-suited for systems with state and input constraints such as MLD and PWA systems. For systems without constraints such as the switched system (1.3), methods based on dynamic programming and particularly DREs seem preferable from a computational complexity perspective due to permitting, at least in part, analytical solutions.

Contributions

A general framework for finite-horizon control and scheduling (FHCS) of switched systems (1.3) unifying and generalizing previous results [ZH08, LR02, LR06, Ran06] is presented in Chapter 2. The solution of Problem 1.9 based on dynamic programming is derived in Section 2.2 and a characterization of the value function is given in Lemma 2.3. Complexity reduction via pruning is addressed in Section 2.3. A necessary and sufficient pruning criterion is formulated in Theorem 2.4. This pruning criterion is, however, computationally intractable. Only an evaluation based on gridding the state space is practicable. Therefore, two sufficient pruning criteria are proposed in Theorems 2.5 and 2.6 together with Algorithms 2.1 and 2.2 for a numerical evaluation. The solution of Problem 1.9 based on relaxed dynamic programming is deduced in Section 2.4. A necessary and sufficient relaxation criterion is given in Theorem 2.8. This relaxation criterion is computationally intractable. Therefore, a sufficient domination criterion is proposed in Theorem 2.9 and based on this Algorithm 2.3 is formulated for relaxed pruning. All criteria are expressed in terms of LMIs. The properties of the methods are illustrated by numerical examples, the effectiveness by a numerical study.

Solutions for a finite time horizon are appropriate for special tracking problems but not for regulation problems where usually solutions for an infinite time horizon are required.

1.4.3 Infinite-Horizon Control and Scheduling

The solution of Problem 1.9 for an infinite time horizon $N = \infty$ can again be deduced from linear-quadratic control theory for discrete-time linear time-varying systems. The stability of the resulting closed-loop system is strongly related to the uniform complete reachability and observability of the switched system (1.3) considering the output vector $\boldsymbol{y}(k) = \boldsymbol{Q}_{1j(k)}^{1/2}\boldsymbol{x}(k)$ as outlined in [KS72, Theorems 6.29 and 6.30]. Uniformity relates, as for stability, to the independence of the initial time. Criteria for checking uniform complete reachability and observability of discrete-time linear time-varying systems have been formulated in terms of Gramians in [KS72, Sections 6.2.7 and 6.2.8]. Though, no computational methods are available for evaluating these criteria for switched systems. Only very few results on the reachability and observability of switched systems exist. In [BE03] it has been shown that the reachability and observability of a discrete-time switched linear system is decidable and criteria for checking the reachability and the observability have been given. Unfortunately, these criteria are very conservative even for small $|\mathbb{M}|$ and n.

For solving Problem 1.9 for an infinite time horizon based on dynamic programming or relaxed dynamic programming using a value iteration, the convergence of the value iteration and coincidence of the limiting value function and the optimal cost function are crucial. The value iteration may furthermore become computationally intractable before convergence is achieved. General criteria for the convergence of the value iteration and relaxed value iteration have been given in [Ran06, Proposition 2, Theorem 1]. Furthermore, general criteria for the coincidence of the limiting value function and the optimal cost function have been indicated in [Ber01, Propositions 3.1.5–3.1.7]. However, no computational methods are available for checking these criteria for switched systems. For Problem 1.9 the convergence of the value function and coincidence of the limiting value function and the optimal cost function are guaranteed if at least one subsystem $(\boldsymbol{A}_{j(k)}, \boldsymbol{B}_{j(k)})$ is stabilizable and $\boldsymbol{Q}_{1j(k)} \succ \boldsymbol{0}$ for all $j(k) \in \mathbb{M}$ as shown in [ZHA09, Theorems 3 and 4]. More general results are, unfortunately, not known.

Optimal control of switched systems and PWA systems for an infinite time horizon has rarely been addressed in the literature. In [ZHA09], it is has been shown that the solution of Problem 1.9 for an infinite time horizon is exponentially stabilizing if at least one subsystem $(\boldsymbol{A}_{j(k)}, \boldsymbol{B}_{j(k)})$ is stabilizable and $\boldsymbol{Q}_{1j(k)} \succ \boldsymbol{0}$ for all $j(k) \in \mathbb{M}$. Unfortunately, no more general results are known. In [BCM06], optimal control of PWA systems for an infinite-time horizon and a linear cost function has been studied. Stability of the resulting closed-loop system has been proved assuming stabilizability of the PWA system and full rank of the state weighting matrix. Though, the proof is specific to the linear cost function.

Despite these important results the solution of Problem 1.9 for an infinite time horizon clearly requires further fundamental research. Alternatively, the infinite-horizon solution may be approximated by a receding-horizon solution or a periodic solution.

1.4.4 Receding-Horizon Control and Scheduling

Receding-horizon control (RHC) or model predictive control (MPC) is widely applied for optimal control of linear systems with constraints, nonlinear systems and recently also MLD and PWA systems. For such systems, an optimization problem for an infinite time horizon can only be solved in special cases. From this perspective, RHC may be considered as an approximation of infinite-horizon control [GR08]. Various surveys [ML99, MRRS00, QB03, etc.] and books [BGW90, Mac02, CB04, KH05, RM09, etc.] on RHC addressing both theoretical and practical aspects have been published in recent years, underlining the large interest both in academia and industry.

The basic idea of RHC consists in solving an optimization problem for a finite time horizon at each time instant with the current state as the initial state. The first element of the solution is applied to the plant; then the procedure is repeated at the next time instant, leading to a receding time horizon as illustrated in Figure 1.2.

The main ingredients of RHC are a model to predict the state vectors $\boldsymbol{x}(k+i+1)$ resulting for the control vectors $\boldsymbol{u}(k+i)$ for $i = 0, \ldots, N-1$ over a prediction horizon N, an optimization problem to determine the optimal control vectors $\boldsymbol{u}^*(k+i)$ for $i = 0, \ldots, N-1$ and the receding horizon principle consisting in applying the first element $\boldsymbol{u}^*(k)$ of the optimal control vector at each time instant k and reoptimizing at

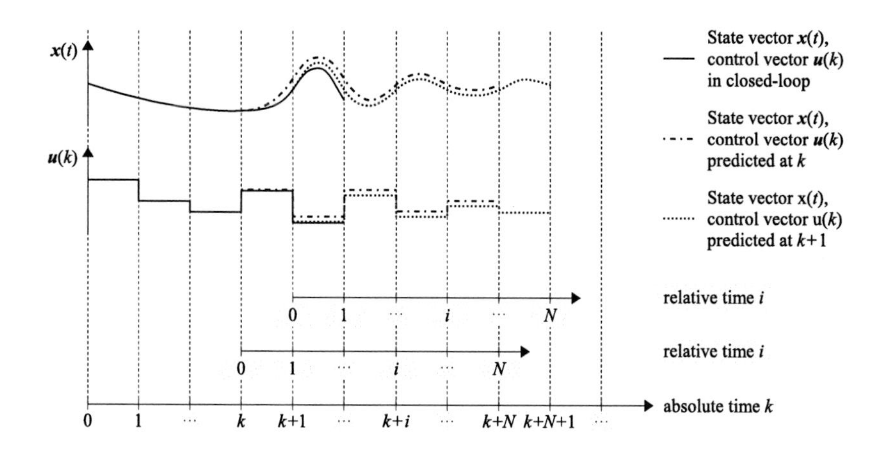

Figure 1.2: Principle of RHC

the next time instant.

The optimization problem solved at each time instant k can be formalized as

$$\begin{aligned} &\min_{\boldsymbol{u}(k),\dots,\boldsymbol{u}(k+N-1)} J_N(\boldsymbol{x}(k)) \\ &\text{subject to } \boldsymbol{x}(k+i+1) = \boldsymbol{f}(\boldsymbol{x}(k+i),\boldsymbol{u}(k+i),k),\ i=0,\dots,N-1 \end{aligned} \tag{1.27}$$

with

$$J_N(\boldsymbol{x}(k)) = F(\boldsymbol{x}(k+N)) + \sum_{i=0}^{N-1} \ell(\boldsymbol{x}(k+i),\boldsymbol{u}(k+i),k+i) \tag{1.28}$$

where $F(\boldsymbol{x}(k+N))$ is the terminal cost and $\ell(\boldsymbol{x}(k+i),\boldsymbol{u}(k+i),k+i)$ is the step cost. Furthermore, k denotes the current time instant, i.e. the absolute time, and i the time instant within the prediction horizon, i.e. the relative time, as illustrated in Figure 1.2. The state vectors $\boldsymbol{x}(k+i)$ and control vectors $\boldsymbol{u}(k+i)$ are often subject to constraints, i.e.

$$\begin{aligned} \boldsymbol{x}(k+i) &\in \mathbb{X} \subseteq \mathbb{R}^{\mathrm{n}} \quad \forall i=0,\dots,N, \\ \boldsymbol{u}(k+i) &\in \mathbb{U} \subseteq \mathbb{R}^{\mathrm{m}} \quad \forall i=0,\dots,N-1. \end{aligned} \tag{1.29}$$

Solving the optimization problem at each time instant can be very time-consuming, which generally limits the application of RHC to plants with slow dynamics or low-order plants. The solution of the optimization problem for linear time-invariant systems with polyhedral state and input constraints and PWA systems can also be expressed explicitly as a PWA state feedback control law

$$\boldsymbol{u}(k) = \boldsymbol{K}_{l(k)}\boldsymbol{x}(k) + \boldsymbol{g}_{\mathrm{C}l(k)} \quad \text{for} \quad \boldsymbol{x}(k) \in \mathcal{X}_{l(k)}. \tag{1.30}$$

The regions $\mathcal{X}_{l(k)}$, $l(k) \in \mathbb{L} = \{1, \dots, L\}$ with $\bigcup_{l(k) \in \mathbb{L}} \mathcal{X}_{l(k)} = \mathbb{X}$ are polyhedral for linear time-invariant systems with linear or quadratic cost functions and for PWA systems with linear cost functions. For PWA systems with quadratic cost functions complex non-convex regions are obtained. Several methods for computing the regions $\mathcal{X}_{l(k)}$, the feedback matrices $\boldsymbol{K}_{l(k)}$ and the feedback vectors $\boldsymbol{g}_{Cl(k)}$ based on multi-parametric programming [BMDP02] for linear time-invariant systems and multi-parametric mixed integer programming [Bor03] or dynamic programming combined with multi-parametric programming [BBBM05] for PWA systems have been proposed in recent years, refer to [AB09] for a survey. The regions and feedback matrices and vectors can be computed offline and stored. The online computation reduces to a region membership test. This so-called explicit RHC is therefore also applicable for plants with fast dynamics.

Closed-loop stability is not guaranteed inherently under RHC. Various methods for ensuring closed-loop stability have been proposed, see [MRRS00] for an excellent overview. Closed-loop stability can be ensured a priori, i.e. before solving the optimization problem, by enforcing that the cost function $J_N(\boldsymbol{x}(k))$ is a Lyapunov function of the closed-loop system. This can be achieved via a terminal state constraint $\boldsymbol{x}(k+N) = \boldsymbol{0}$ [KG88], a terminal cost $F(\boldsymbol{x}(k+N))$ [BGW90, RM93], a terminal constraint set $\boldsymbol{x}(k+N) \in \mathbb{X}_T$ [MM93, CM96, SR98] or a combination of these. The terminal cost can be interpreted as an approximation of the cost beyond the prediction horizon, i.e. from $k = N+1$ to $k = \infty$. In this way an infinite time horizon is emulated whereby closed-loop stability is guaranteed under appropriate reachability and observability conditions as outlined in Section 1.4.3. These concepts have also been transcribed for RHC of MLD and PWA systems [BM99, MR03, GKBM05, LHWB06, Laz06]. Imposing constraints on the cost function $J_N(\boldsymbol{x}(k))$ may considerably change the optimization problem and may for non-linear and PWA systems even impair robustness as shown in [GMTT04, LHT09] and [Laz06, Section 4.2]. Therefore, methods for guaranteeing closed-loop stability without imposing constraints on the cost function have been intensely studied in recent years, e.g. in [NP97, PN00, GMTT05, GR08, Grü09]. These methods aim at specifying the prediction horizon N for which closed-loop stability is ensured. Closed-loop stability can also be analyzed a posteriori, i.e. after solving the optimization problem. On the one hand, it can be checked whether the cost function $J_N(\boldsymbol{x}(k))$ is a Lyapunov function of the closed-loop system. This is the cornerstone for ensuring closed-loop stability without imposing constraints on the cost function [NP97, PN00, GMTT05, GR08, Grü09]. On the other hand, an "artificial" Lyapunov function can be constructed such as a PWA or PWQ Lyapunov function. This is frequently considered for explicit RHC and RHC of PWA systems [BGLM05].

Remarkably, while RHC of PWA systems has been intensely studied, RHC has not been considered for other switched systems and particularly not for Problem 1.9.

Contributions

A general framework for receding-horizon control and scheduling (RHCS) of switched systems (1.3) is proposed in Chapter 3. The basic idea of RHCS consists in applying the solution of Problem 1.9 derived in Chapter 2 based on dynamic programming and relaxed dynamic programming according to the receding horizon principle. The resulting RHCS strategy can be expressed explicitly as a piecewise linear (PWL) state feedback control law defined over regions which are characterized by quadratic forms as shown in Section 3.2 (Theorem 3.2). Several methods for assessing closed-loop stability under the RHCS strategy are presented in Section 3.3: An a priori stability condition based on a terminal weighting matrix $\boldsymbol{Q}_0$ is derived in Section 3.3.1 (Theorems 3.3 and 3.4). This condition is only applicable if Problem 1.9 is solved without relaxation. Various a posteriori stability criteria are devised in Section 3.3.2 based on constructing a PWQ Lyapunov function (Theorems 3.6 and 3.8) or on using the value function as a candidate Lyapunov function (Theorem 3.9). These criteria are also applicable if Problem 1.9 is solved with relaxation. PWQ Lyapunov functions are widely considered for analyzing the stability of PWA systems defined over polyhedral regions [JR98, MFM00, FCMM02] as outlined in Section 1.3.3. By contrast, PWQ Lyapunov functions are employed here for PWL systems defined over regions which are characterized by quadratic forms. A method for handling such regions is formulated in Lemma 3.7. This method allows including the regions in the PWQ Lyapunov function whereby conservatism is alleviated. Regions characterized by quadratic forms frequently arise in optimal control of switched systems and PWA systems under a quadratic cost function. Therefore, Lemma 3.7 and Theorems 3.6 and 3.8 are also relevant under a wider scope. Based on Lemma 3.7 additionally a region-reachability criterion is deduced in Section 3.4 (Theorem 3.11). All conditions and criteria are formulated in terms of LMIs. The methods are illustrated by numerical examples, the effectiveness by numerical studies.

1.4.5 Periodic Control and Scheduling

The solution of Problem 1.9 for an infinite time horizon $N = \infty$ and a p-periodic switching sequence $j(k) = j(k+p)\ \forall k \in \mathbb{N}_0$ can be deduced from periodic control theory [BCD91, BC09]: The p-periodic switching sequence $j(k)$ is fixed; then the optimal control sequence $\boldsymbol{u}^*(k)$ is determined from a discrete-time periodic Riccati equation (DPRE). This is repeated for all possible p-periodic switching sequences $j(k) \in \mathbb{M}^p$. The optimal p-periodic switching sequence $j^*(k)$ is determined by searching the switching sequence which yields the minimum cost among all possible p-periodic switching sequences $j(k) \in \mathbb{M}^p$. This procedure is formalized in Section 4.1.3. Similar concepts have been proposed in [RS00, RS04, BÇH06, BÇH09, LHB09] for control and scheduling codesign of networked embedded control systems. Scheduling is performed completely offline. The optimal p-periodic switching sequence $j^*(k)$ is therefore not adapted under disturbances which may considerably impair the performance. Moreover, the degrees of freedom for the optimization are reduced by imposing periodicity which may further degrade the performance. Initial ideas to gain adaptivity have been reported in [BÇH06, BÇH09]: An optimal p-periodic switching sequence $j^*(k)$ and p-periodic state feedback control law are first determined offline and then adapted online by reoptimizing for the current state over all cyclic shifts of $j^*(k)$. This strategy is denoted as optimal pointer placement (OPP). Remarkably, closed-loop stability is guaranteed inherently under the OPP strategy. Further developing these ideas is surely very promising.

Contributions

A general framework for periodic control and scheduling (PCS) of switched systems (1.3) is presented in Chapter 4. Periodic control theory is reviewed in Section 4.1.2. The solution of Problem 1.9 for an infinite time horizon and a p-periodic switching sequence is then derived based on periodic control theory and exhaustive search in Section 4.1.3 (Theorem 4.11 and Algorithm 4.1), leading to a periodic control and offline scheduling (PCS_{off}) strategy. Further, a periodic control and online scheduling (PCS_{on}) strategy is proposed in Section 4.2: The solutions of Problem 1.9 for an infinite time horizon and all possible p-periodic switching sequences $j(k) \in \mathbb{M}^p$ are computed offline and stored. The solution yielding the minimum cost for the current state $\boldsymbol{x}(k)$ is then selected online at each time instant k. The resulting PCS_{on} strategy is therefore adaptive under disturbances and additional degrees of freedom for the optimization are gained. The PCS_{on} strategy can be expressed explicitly as a PWL state feedback control law defined over regions which are characterized by quadratic forms as pointed out in Section 4.2.2 (Theorem 4.13). Closed-loop stability is guaranteed inherently as proved in Section 4.2.3 (Theorem 4.14). The number of p-periodic switching sequences is determined by $|\mathbb{M}|^p$ and therefore increases exponentially with the period p. Selecting the solution yielding the minimum cost online is consequently only feasible for small $|\mathbb{M}|$ and p. Three methods for reducing the online complexity are presented in Section 4.2.4. First, the concept of relaxation is utilized for reducing the online complexity, yielding a relaxed periodic control and online scheduling ($RPCS_{on}$) strategy. Closed-loop stability under the $RPCS_{on}$ strategy can always be guaranteed under mild conditions as shown in Theorem 4.21. Second, a heuristic based on preserving the distribution of the p-periodic switching sequences is proposed, leading to a heuristic periodic control and online scheduling ($HPCS_{on}$) strategy. Closed-loop stability is guaranteed inherently under the $HPCS_{on}$ strategy. Third, the OPP strategy is presented as a special case of the PCS_{on} strategy. Notably, the cost resulting under the PCS_{on}, $RPCS_{on}$, $HPCS_{on}$ and OPP strategy is always less than or equal to the cost resulting under the PCS_{off} strategy. The effectiveness of the methods is shown by comprehensive numerical studies.

2 Finite-Horizon Control and Scheduling

2.1 Problem Formulation

Consider the quadratic discrete-time cost function

$$J_N = \boldsymbol{x}^T(N)\boldsymbol{Q}_0\boldsymbol{x}(N) + \sum_{k=0}^{N-1} \left[\boldsymbol{x}^T(k)\boldsymbol{Q}_{1j(k)}\boldsymbol{x}(k) + \boldsymbol{u}^T(k)\boldsymbol{Q}_{2j(k)}\boldsymbol{u}(k)\right] \tag{2.1}$$

where $\boldsymbol{Q}_0 \succeq 0$ is a symmetric terminal weighting matrix, $\boldsymbol{Q}_{1j(k)} \succeq 0$ and $\boldsymbol{Q}_{2j(k)} \succ 0$ with $j(k) \in \mathbb{M}$ are symmetric switched weighting matrices and N is a finite time horizon.

Problem 2.1 *For the switched system (1.3) find a control sequence* $\boldsymbol{u}^*(k)$ *and a switching sequence* $j^*(k)$ *with* $k = 0, \ldots, N-1$ *such that the cost function (2.1) is minimized, i.e.*

$$\min_{\substack{\boldsymbol{u}(0),\ldots,\boldsymbol{u}(N-1)\\ j(0),\ldots,j(N-1)}} J_N \qquad \text{subject to (1.3).} \tag{2.2}$$

2.2 Dynamic Programming Solution

Theorem 2.2 *Consider the difference Riccati equation (DRE)*

$$\boldsymbol{P}_{l(k)} = (\boldsymbol{A}_{j(k)} - \boldsymbol{B}_{j(k)}\boldsymbol{K}_{l(k)})^T\boldsymbol{P}_{l(k+1)}(\boldsymbol{A}_{j(k)} - \boldsymbol{B}_{j(k)}\boldsymbol{K}_{l(k)}) + \boldsymbol{Q}_{1j(k)} + \boldsymbol{K}_{l(k)}^T\boldsymbol{Q}_{2j(k)}\boldsymbol{K}_{l(k)} \tag{2.3}$$

with the terminal condition $\boldsymbol{P}_{l(N)} = \boldsymbol{Q}_0$. *The controller index* $l(k) \in \mathbb{L}_k = \{1, \ldots, L_k\}$ *with* $L_k = M^{N-k}$ *indexes the DRE solutions* $\boldsymbol{P}_{l(k)}$ *and the associated switched feedback matrices* $\boldsymbol{K}_{l(k)}$ *at time instant* k. *The matrices* $\boldsymbol{P}_{l(k)}$ *result from the matrices* $\boldsymbol{P}_{l(k+1)}$ *by evaluating (2.3) for all combinations of the current switching index* $j(k) \in \mathbb{M}$ *and the subsequent controller index* $l(k+1) \in \mathbb{L}_{k+1}$ *as illustrated by the switching tree in Figure 2.1.*

Problem 2.1 is then solved with minimum cost

$$J_N^* = \min_{l \in \mathbb{L}_0} \boldsymbol{x}_0^T\boldsymbol{P}_l\boldsymbol{x}_0 \tag{2.4}$$

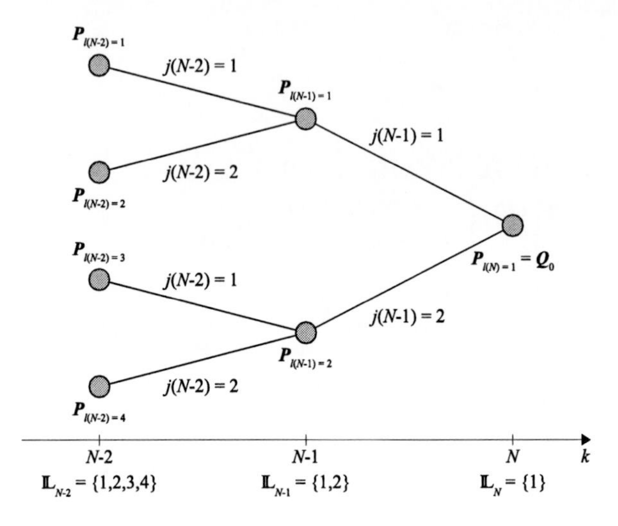

Figure 2.1: Switching tree for $M = 2$

by the optimal controller sequence starting with

$$l^*(0) = \arg \min_{l \in \mathbb{L}_0} \boldsymbol{x}_0^T \boldsymbol{P}_l \boldsymbol{x}_0, \tag{2.5}$$

the optimal switching sequence starting with

$$j^*(0) = \sigma(l^*(0)) \tag{2.6}$$

and the optimal switched state feedback control law

$$\boldsymbol{u}^*(k) = -\boldsymbol{K}_{l^*(k)} \boldsymbol{x}(k) \tag{2.7}$$

with the switched feedback matrix

$$\boldsymbol{K}_{l^*(k)} = (\boldsymbol{Q}_{2j^*(k)} + \boldsymbol{B}_{j^*(k)}^T \boldsymbol{P}_{l^*(k+1)} \boldsymbol{B}_{j^*(k)})^{-1} \boldsymbol{B}_{j^*(k)}^T \boldsymbol{P}_{l^*(k+1)} \boldsymbol{A}_{j^*(k)}. \tag{2.8}$$

The remaining controller sequence $l^*(k)$ and switching sequence $j^*(k) = \sigma(l^*(k))$ follow for $k = 1, \ldots, N-1$ from iterating the switching tree backwards, where the switching function σ follows by construction.

Remark 2.1. The switching function $\sigma : \mathbb{L}_k \to \mathbb{M}$ which maps the controller index $l(k)$ onto the switching index $j(k)$ results from iterating the switching tree backwards. Note that the switching function is not injective since some $l(k)$ may be mapped onto the same $j(k)$ but surjective since each $j(k) \in \mathbb{M}$ is taken as can be seen in Figure 2.1.

PROOF. The proof of Theorem 2.2 is based on dynamic programming. Introduce the value function $V^*_{N-k}(k) = V^*_{N-k}(\boldsymbol{x}(k))$ defined by

$$V^*_{N-k}(k) = \min_{\substack{\boldsymbol{u}(k),\dots,\boldsymbol{u}(N-1)\\ j(k),\dots,j(N-1)}} \left\{ \boldsymbol{x}^T(N)\boldsymbol{Q}_0\boldsymbol{x}(N) + \sum_{i=k}^{N-1} \ell(i) \right\} \tag{2.9}$$

with the non-negative step cost $\ell(k) = \ell(\boldsymbol{x}(k), \boldsymbol{u}(k), j(k))$ defined by

$$\ell(k) = \boldsymbol{x}^T(k)\boldsymbol{Q}_{1j(k)}\boldsymbol{x}(k) + \boldsymbol{u}^T(k)\boldsymbol{Q}_{2j(k)}\boldsymbol{u}(k). \tag{2.10}$$

The value function describes the minimum cost-to-go from k to N, where the subindex $N-k$ specifies the steps-to-go. It can also be written as the Bellman equation [Ber05a]

$$\begin{aligned} V^*_{N-k}(k) &= \min_{\substack{\boldsymbol{u}(k)\\ j(k)}} \left\{ \ell(k) + V^*_{N-(k+1)}(k+1) \right\} \\ V^*_0(N) &= \boldsymbol{x}^T(N)\boldsymbol{Q}_0\boldsymbol{x}(N). \end{aligned} \tag{2.11}$$

Following the principle of dynamic programming, the value function is iterated backwards from $k = N$ to $k = 0$.

For the last optimization step $k = N$ the value function is given by

$$V^*_0(N) = \boldsymbol{x}^T(N)\boldsymbol{P}_{l(N)}\boldsymbol{x}(N) \tag{2.12}$$

with $\boldsymbol{P}_{l(N)} = \boldsymbol{Q}_0$ and $l(N) \in \mathbb{L}_N = \{1\}$.

For the second to last optimization step $k = N-1$ the value function becomes

$$V^*_1(N-1) = \min_{\substack{\boldsymbol{u}(N-1)\\ j(N-1)}} \left\{ \ell(N-1) + V^*_0(N) \right\}. \tag{2.13}$$

Substituting (2.12) with $\boldsymbol{x}(N) = \boldsymbol{A}_{j(N-1)}\boldsymbol{x}(N-1) + \boldsymbol{B}_{j(N-1)}\boldsymbol{u}(N-1)$ into (2.13) yields

$$\begin{aligned} V^*_1(N-1) = &\min_{\substack{\boldsymbol{u}(N-1)\\ j(N-1)}} \Big\{ \ell(N-1) + \left(\boldsymbol{A}_{j(N-1)}\boldsymbol{x}(N-1) + \boldsymbol{B}_{j(N-1)}\boldsymbol{u}(N-1)\right)^T \boldsymbol{P}_{l(N)} \\ &\qquad \left(\boldsymbol{A}_{j(N-1)}\boldsymbol{x}(N-1) + \boldsymbol{B}_{j(N-1)}\boldsymbol{u}(N-1)\right)\Big\} \\ = &\min_{\substack{\boldsymbol{u}(N-1)\\ j(N-1)}} f(\boldsymbol{x}(N-1), \boldsymbol{u}(N-1), j(N-1)). \end{aligned} \tag{2.14}$$

The minimization of $f(\boldsymbol{x}(N-1), \boldsymbol{u}(N-1), j(N-1))$ with respect to $\boldsymbol{u}(N-1)$ is an unconstrained quadratic programming problem which can be solved via

$$\begin{aligned} \frac{\partial f}{\partial \boldsymbol{u}(N-1)} =& 2\left(\boldsymbol{Q}_{2j(N-1)} + \boldsymbol{B}^T_{j(N-1)}\boldsymbol{P}_{l(N)}\boldsymbol{B}_{j(N-1)}\right)\boldsymbol{u}(N-1) + \\ & 2\boldsymbol{B}^T_{j(N-1)}\boldsymbol{P}_{l(N)}\boldsymbol{A}_{j(N-1)}\boldsymbol{x}(N-1) \\ \stackrel{!}{=}& 0. \end{aligned} \tag{2.15}$$

The optimal control law follows as

$$\begin{aligned} \boldsymbol{u}^*(N-1) &= -\left(\boldsymbol{Q}_{2j(N-1)} + \boldsymbol{B}_{j(N-1)}^T \boldsymbol{P}_{l(N)} \boldsymbol{B}_{j(N-1)}\right)^{-1} \boldsymbol{B}_{j(N-1)}^T \boldsymbol{P}_{l(N)} \boldsymbol{A}_{j(N-1)} \boldsymbol{x}(N-1) \\ &= -\boldsymbol{K}_{l(N-1)}^* \boldsymbol{x}(N-1). \end{aligned} \tag{2.16}$$

Substituting (2.16) into (2.14) yields

$$\begin{aligned} V_1^*(N-1) = \min_{j(N-1)} \boldsymbol{x}^T(N-1) \Big\{ & \left(\boldsymbol{A}_{j(N-1)} - \boldsymbol{B}_{j(N-1)} \boldsymbol{K}_{l(N-1)}^*\right)^T \boldsymbol{P}_{l(N)} \\ & \left(\boldsymbol{A}_{j(N-1)} - \boldsymbol{B}_{j(N-1)} \boldsymbol{K}_{l(N-1)}^*\right) + \\ & \boldsymbol{Q}_{1j(N-1)} + \boldsymbol{K}^{*T}_{l(N-1)} \boldsymbol{Q}_{2j(N-1)} \boldsymbol{K}_{l(N-1)}^* \Big\} \boldsymbol{x}(N-1) \\ = \min_{j(N-1)} g(\boldsymbol{x}(N-1), j(N-1)). \end{aligned} \tag{2.17}$$

The minimization of $g(\boldsymbol{x}(N-1), j(N-1))$ with respect to $j(N-1) \in \mathbb{M}$ is a combinatorial optimization problem which can be solved via explicit enumeration. This explicit enumeration is inherently included in the DRE (2.3) via the controller index $l(N-1) \in \mathbb{L}_{N-1} = \{1, \dots, M\}$. Thus, using (2.3) yields

$$V_1^*(N-1) = \min_{l(N-1) \in \mathbb{L}_{N-1}} \boldsymbol{x}^T(N-1) \boldsymbol{P}_{l(N-1)} \boldsymbol{x}(N-1). \tag{2.18}$$

For $k = N-2$ the value function corresponds to (2.13) with shifted time instant. Continuing the backward iteration, for $k = 0$ finally

$$J_N^* = V_N^*(0) = \min_{l(0) \in \mathbb{L}_0} \boldsymbol{x}_0^T \boldsymbol{P}_{l(0)} \boldsymbol{x}_0 \tag{2.19}$$

is obtained. Thus (2.3) to (2.8) are proved where $l(0) \in \mathbb{L}_0$ and $\boldsymbol{K}_{l^*(k)}^*$ are shortly written as $l \in \mathbb{L}_0$ and $\boldsymbol{K}_{l^*(k)}$. This notation will also be used in the following. □

Remark 2.2. For the existence of $\boldsymbol{K}_{l^*(k)}$ it is sufficient to require that $\boldsymbol{Q}_{2j(k)} \succeq \boldsymbol{0}$ and $\boldsymbol{Q}_{2j(k)} + \boldsymbol{B}_{j(k)}^T \boldsymbol{P}_{l(k+1)} \boldsymbol{B}_{j(k)}$ is regular for all $k \in \{0, \dots, N-1\}$. The latter condition can however usually not be ensured in advance.

Remark 2.3. Throughout the thesis, a distinction between a specific controller index and an arbitrary controller index is often required. A specific controller index will be denoted by m or n, an arbitrary controller index by l. Similarly, a distinction between a specific switching index or sequence and an arbitrary switching index or sequence is often required. A specific switching index or sequence will be denoted by v or w, an arbitrary switching sequence or index by j. This notation will be frequently used and should therefore be borne in mind.

The value function

$$V_{N-k}^*(k) = \min_{l \in \mathbb{L}_k} \boldsymbol{x}^T(k) \boldsymbol{P}_l \boldsymbol{x}(k) \tag{2.20}$$

which is the pointwise minimum of quadratic forms has some important properties:

Lemma 2.3 *The value function* $V^*_{N-k}(\boldsymbol{x}(k))$ *is homogeneous of degree 2 and even, i.e.*

$$V^*_{N-k}(\gamma\boldsymbol{x}(k)) = \gamma^2 V^*_{N-k}(\boldsymbol{x}(k)) \qquad \forall \boldsymbol{x}(k) \in \mathbb{R}^{\mathrm{n}} \quad \forall \gamma \in \mathbb{R} \tag{2.21a}$$
$$V^*_{N-k}(-\boldsymbol{x}(k)) = V^*_{N-k}(\boldsymbol{x}(k)) \qquad \forall \boldsymbol{x}(k) \in \mathbb{R}^{\mathrm{n}}. \tag{2.21b}$$

PROOF. The proof follows from

$$V^*_{N-k}(\gamma\boldsymbol{x}(k)) = \min_{l\in\mathbb{L}_k}(\gamma\boldsymbol{x}(k))^T\boldsymbol{P}_l(\gamma\boldsymbol{x}(k)) = \gamma^2\min_{l\in\mathbb{L}_k}\boldsymbol{x}^T(k)\boldsymbol{P}_l\boldsymbol{x}(k) = \gamma^2 V^*_{N-k}(\boldsymbol{x}(k)) \tag{2.22a}$$
$$V^*_{N-k}(-\boldsymbol{x}(k)) = \min_{l\in\mathbb{L}_k}(-\boldsymbol{x}(k))^T\boldsymbol{P}_l(-\boldsymbol{x}(k)) = \min_{l\in\mathbb{L}_k}\boldsymbol{x}^T(k)\boldsymbol{P}_l\boldsymbol{x}(k) = V^*_{N-k}(\boldsymbol{x}(k)). \tag{2.22b}$$

□

The homogeneity of the value function implies that in view of (2.5) and (2.6) the optimal controller index $l^*(k)$ and the optimal switching index $j^*(k)$ are radially invariant, i.e. $l^*(k)$ and $j^*(k)$ are constant for all $\boldsymbol{x}(k)$ having the same radial direction. Furthermore, the homogeneity and evenness entail that operations on the value function concerning $l^*(k)$ and $j^*(k)$ may be restricted to a unit semi-hypersphere, e.g. $\mathcal{S}_{\mathrm{semi}} = \{\boldsymbol{x} \in \mathbb{R}^{\mathrm{n}} | \, \|\boldsymbol{x}\|_2 = 1, x_{\mathrm{n}} \geq 0\}$. This may in particular be exploited for operations based on gridding of the state space $\mathbb{R}^{\mathrm{n}}$.

The properties of the value function are illustrated by the following

Example 2.1 Consider the switched system (1.3) with the system and input matrices

$$\boldsymbol{A}_1 = \begin{pmatrix} 0 & 1 \\ -0.8 & 2.4 \end{pmatrix}, \ \boldsymbol{A}_2 = \begin{pmatrix} 0 & 1 \\ -1.8 & 3.6 \end{pmatrix}, \ \boldsymbol{A}_3 = \begin{pmatrix} 0 & 1 \\ -0.56 & 1.8 \end{pmatrix}, \ \boldsymbol{A}_4 = \begin{pmatrix} 0 & 1 \\ -8 & 6 \end{pmatrix}$$
$$\boldsymbol{B}_1 = \boldsymbol{B}_2 = \boldsymbol{B}_3 = \boldsymbol{B}_4 = \begin{pmatrix} 0 \\ 1 \end{pmatrix} \tag{2.23}$$

and the cost function (2.1) with the time horizon $N = 10$ and the weighting matrices

$$\begin{aligned} \boldsymbol{Q}_0 = \boldsymbol{Q}_{11} = \boldsymbol{Q}_{12} = \boldsymbol{Q}_{13} = \boldsymbol{Q}_{14} = \boldsymbol{I} \\ \boldsymbol{Q}_{21} = \boldsymbol{Q}_{22} = \boldsymbol{Q}_{23} = \boldsymbol{Q}_{24} = 10, \end{aligned} \tag{2.24}$$

whereby Problem 2.1 is characterized completely. Note that the subsystems $(\boldsymbol{A}_{j(k)}, \boldsymbol{B}_{j(k)})$ with $j(k) \in \mathbb{M} = \{1, \ldots, M\}$, $M = 4$ are unstable but stabilizable.

Considering optimization step $k = 9$, the value function $V_1(\boldsymbol{x}(9))$ and the quadratic forms $\boldsymbol{x}^T(9)\boldsymbol{P}_{l(9)}\boldsymbol{x}(9)$ associated to the DRE solutions $\boldsymbol{P}_{l(9)}$ with $l \in \mathbb{L}_9 = \{1, \ldots, L_9\}$, $L_9 = 4^{10-9} = 4$ which constitute the value function are plotted in Figure 2.2 (left) on the unit semi-circle $\boldsymbol{x}(9) = \begin{pmatrix} \cos\varphi & \sin\varphi \end{pmatrix}^T$ with $\varphi \in [0, \pi]$. Clearly, the minimum quadratic form and therefore the optimal controller index $l^*(9)$ and optimal switching index $j^*(9)$ vary over φ. The value function on the unit semi-circle for the optimization steps $k = 10, \ldots, 4$ is depicted in Figure 2.2 (right). Obviously, the value function converges over the optimization steps as claimed in [ZHA09, Theorem 3] since $(\boldsymbol{A}_{j(k)}, \boldsymbol{B}_{j(k)})$ is stabilizable and $\boldsymbol{Q}_{1j(k)} \succ \boldsymbol{0}$ for all $j(k) \in \mathbb{M}$. □

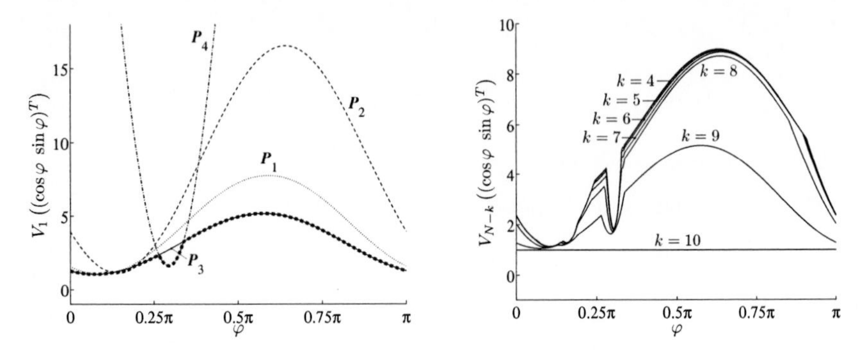

Figure 2.2: Value function (bold and dashed) and quadratic forms (non-bold and solid, dashed, dotted or dash-dotted) for optimization step $k = 9$ (left) and value function for optimization steps $k = 10, \dots, 4$ (right) on the unit semi-circle

The number of DRE solutions $\boldsymbol{P}_{l(k)}$ given by $L_k = |\mathbb{L}_k| = M^{N-k}$ grows exponentially with the time horizon N. Therefore, the solution of Problem 2.1 rapidly becomes computationally intractable. In the following section a pruning method is introduced to cope with computational complexity.

2.3 Complexity Reduction via Pruning

Theorem 2.4 (Pruning Criterion) *Iff for each $\boldsymbol{x}(k) \in \mathbb{R}^n$ there exists at least one $l \neq m$ with $l, m \in \mathbb{L}_k$ such that*

$$\boldsymbol{x}^T(k)\boldsymbol{P}_m\boldsymbol{x}(k) \geq \boldsymbol{x}^T(k)\boldsymbol{P}_l\boldsymbol{x}(k), \tag{2.25}$$

then the node $\boldsymbol{P}_m$ can be pruned from the switching tree.

PROOF. Reconsider the value function (2.20). If for each state $\boldsymbol{x}(k) \in \mathbb{R}^n$ the value resulting from $\boldsymbol{x}^T(k)\boldsymbol{P}_m\boldsymbol{x}(k)$ exceeds the value resulting from $\min_{l\in\mathbb{L}_k\setminus\{m\}} \boldsymbol{x}^T(k)\boldsymbol{P}_l\boldsymbol{x}(k)$, then the control and switching sequence corresponding to $\boldsymbol{P}_m$ can not be optimal at time instant k and thus the value function (2.20) is equivalent to

$$V^*_{N-k}(k) = \min_{l\in\mathbb{L}_k\setminus\{m\}} \boldsymbol{x}^T(k)\boldsymbol{P}_l\boldsymbol{x}(k) \quad \forall \boldsymbol{x}(k) \in \mathbb{R}^n. \tag{2.26}$$

This is illustrated schematically in Figure 2.3. In this figure the value function is plotted bold and dashed, the quadratic forms constituting the value function are plotted non-bold and dashed, dotted or dash-dotted. The value resulting from $\boldsymbol{x}^T(k)\boldsymbol{P}_2\boldsymbol{x}(k)$ exceeds the value resulting from $\min_{l\in\{1,2,3\}\setminus\{2\}} \boldsymbol{x}^T(k)\boldsymbol{P}_l\boldsymbol{x}(k)$ for each state $\boldsymbol{x}(k) \in \mathbb{R}^n$. Thus, $\boldsymbol{P}_2$ does not contribute to the minimum.

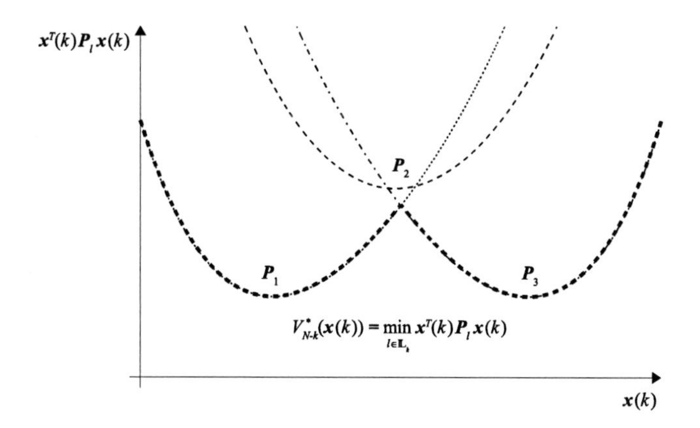

Figure 2.3: Illustration of the Pruning Criterion

According to the Bellman equation (2.11), also at time instants $k-1, \ldots, 0$ the value function can not depend on $\boldsymbol{P}_m$. Hence, the node $\boldsymbol{P}_m$ can be pruned from the switching tree. This is illustrated in Figure 2.4, where $\hat{l}(k) \in \hat{\mathbb{L}}_k$ denotes the controller index after pruning.

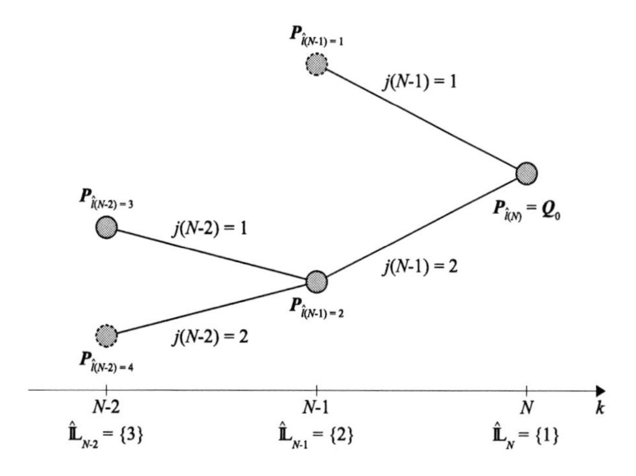

Figure 2.4: Switching tree for $M = 2$ with two nodes pruned (dashed)

□

Remark 2.4. The switching function σ is not surjective anymore after pruning. For example, in Figure 2.4 we have $\sigma(\hat{l}) \neq 1 \in \mathbb{M}$ for any $\hat{l} \in \hat{\mathbb{L}}_{N-1}$.

Remark 2.5. Note that the notation introduced in Remark 2.3 is utilized in Theorem 2.4. Precisely, m denotes the specific controller index under investigation while $l \neq m$ denotes an arbitrary controller index.

The pruning criterion proposed in Theorem 2.4 has to be checked for each $\boldsymbol{x}(k) \in \mathbb{R}^n$ or, more precisely, for each $\boldsymbol{x}(k) \in \mathcal{S}_{\text{semi}}$, see Lemma 2.3 and the subsequent discussion. Since the controller index $l \in \mathbb{L}_k \backslash \{m\}$ fulfilling the condition (2.25) may vary over $\boldsymbol{x}(k) \in \mathbb{R}^n$ or $\boldsymbol{x}(k) \in \mathcal{S}_{\text{semi}}$, this is clearly not possible from a computational perspective. Instead, a common controller index $n \in \mathbb{L}_k \backslash \{m\}$ can be considered for all states $\boldsymbol{x}(k) \in \mathbb{R}^n$ to prune $\boldsymbol{P}_m$, leading to the following computationally tractable pruning criterion which is only sufficient:

Theorem 2.5 (Pruning Criterion Based on a Common n) *If there exists a common $n \neq m$ with $n, m \in \mathbb{L}_k$ such that*

$$\boldsymbol{x}^T(k)\boldsymbol{P}_m\boldsymbol{x}(k) \geq \boldsymbol{x}^T(k)\boldsymbol{P}_n\boldsymbol{x}(k) \quad \forall \boldsymbol{x}(k) \in \mathbb{R}^n \tag{2.27}$$

or equivalently

$$\boldsymbol{P}_m - \boldsymbol{P}_n \succeq \boldsymbol{0}, \tag{2.28}$$

then the node $\boldsymbol{P}_m$ can be pruned from the switching tree.

Proof. Follows immediately from the proof of Theorem 2.4 by setting $l = n =$ const. for all $\boldsymbol{x}(k) \in \mathbb{R}^n$. A schematic illustration is given in Figure 2.5.

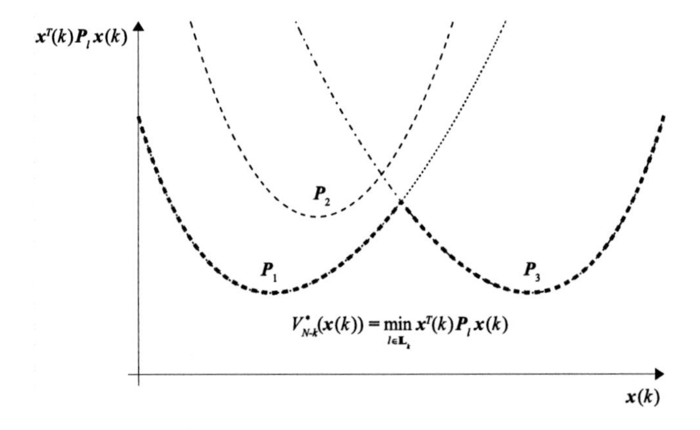

Figure 2.5: Illustration of the Pruning Criterion Based on a Common n

In this figure the value function is plotted bold and dashed, the quadratic forms constituting the value function are plotted non-bold and dashed, dotted or dash-dotted. The value resulting from $\boldsymbol{x}^T(k)\boldsymbol{P}_2\boldsymbol{x}(k)$ exceeds the value resulting from $\boldsymbol{x}^T(k)\boldsymbol{P}_1\boldsymbol{x}(k)$ for all states $\boldsymbol{x}(k) \in \mathbb{R}^n$, i.e. $n = 1$ is a common controller index for $\boldsymbol{P}_2$ or, loosely speaking, $\boldsymbol{P}_2$ is dominated by $\boldsymbol{P}_1$. □

An algorithm implementing pruning according to Theorem 2.5 is given in Algorithm 2.1.

Algorithm 2.1 Pruning Algorithm Mapping $\mathbb{L}_k \to \hat{\mathbb{L}}_k$

```
Input:   P_l, l ∈ 𝕃_k // DRE solutions before pruning, possibly sorted by trace
Output:  P_l̂, l̂ ∈ 𝕃̂_k // DRE solutions after pruning
  𝕃̂_k = {1} // initialize pruned controller index set 𝕃̂_k
  for m = 2 to L_k do
    prunable = false // DRE solution not prunable (initialization)
    for all l̂ ∈ 𝕃̂_k do
      if P_m − P_l̂ ⪰ 0 then
        prunable = true // DRE solution prunable
        break for loop
      end if
    end for
    if prunable = false then
      𝕃̂_k = 𝕃̂_k ∪ {m} // retain controller index
    end if
  end for
```

The algorithm evaluates the condition (2.28) for each DRE solution $\boldsymbol{P}_l$ with $l \in \mathbb{L}_k$. If the condition is fulfilled, the DRE solution is pruned. Otherwise, the DRE solution is retained. For evaluating the condition, only those DRE solutions $\boldsymbol{P}_{\hat{l}}$ with $\hat{l} \in \hat{\mathbb{L}}_k$ are considered which have already proved not prunable. It is therefore essential for computational efficiency to initially evaluate the condition for DRE solutions which are presumably not prunable. This can be achieved by presorting the DRE solutions by trace as proposed in [LR06], i.e. reassigning the controller indices $l \in \mathbb{L}_k$ of $\boldsymbol{P}_l$ such that

$$\operatorname{tr}(\boldsymbol{P}_1) \leq \operatorname{tr}(\boldsymbol{P}_2) \leq \ldots \leq \operatorname{tr}(\boldsymbol{P}_{L_k}), \tag{2.29}$$

which is motivated by the fact that the trace of a matrix corresponds to the expected value of the related quadratic form as shown in Appendix A.1.

The pruning criterion in Theorem 2.5 is based on bounding a DRE solution by a single DRE solution from below. This is quite restrictive in comparison to Theorem 2.4. Instead, a DRE solution can also be bounded from below by a convex combination of DRE solutions using the S-procedure.

Theorem 2.6 (Pruning Criterion Based on the S-Procedure) *If there exist non-negative scalars ξ_l with $l \in \mathbb{L}_k \backslash \{m\}$ such that*

$$\sum_{l \in \mathbb{L}_k \backslash \{m\}} \xi_l = 1 \quad \text{and} \quad \boldsymbol{P}_m \succeq \sum_{l \in \mathbb{L}_k \backslash \{m\}} \xi_l \boldsymbol{P}_l, \tag{2.30}$$

then the node $\boldsymbol{P}_m$ can be pruned.

PROOF. Equation (2.30) can be reformulated as

$$\sum_{l \in \mathbb{L}_k \setminus \{m\}} \xi_l (\boldsymbol{P}_m - \boldsymbol{P}_l) \succeq 0. \tag{2.31}$$

If inequality (2.31) holds for some non-negative scalars ξ_l, then for each $\boldsymbol{x}(k) \in \mathbb{R}^{\mathrm{n}}$ there exists at least one $l \in \mathbb{L}_k \setminus \{m\}$ such that $\boldsymbol{x}^T(k)(\boldsymbol{P}_m - \boldsymbol{P}_l)\boldsymbol{x}(k) \geq 0$. Therefore (2.30) implies the pruning criterion given in Theorem 2.4. The reverse implication does not hold. This approach relates to the S-procedure, see e.g. [Yak77], [BEFB94, Section 2.6.3] and [SP05, Section 12.3.4] for a general treatment. □

Remark 2.6. A similar pruning criterion as in Theorem 2.6 has been proposed in [ZH08, Lemma 3]. However, no proof has been given.

Remark 2.7. The pruning criterion in Theorem 2.6 is equivalent to the pruning criterion in Theorem 2.5 when restricting ξ_l to $\xi_n = 1$ for some $n \in \mathbb{L}_k \setminus \{m\}$ and $\xi_l = 0$ for all other $l \in \mathbb{L}_k \setminus \{m, n\}$ which entails that the former is more general and thus less conservative.

Remark 2.8. The pruning criterion in Theorem 2.6 corresponds to an LMI feasibility problem. Its evaluation is therefore computationally much more demanding than the evaluation of the pruning criterion in Theorem 2.5 which only includes a definiteness test.

An algorithm implementing pruning according to Theorem 2.6 is given in Algorithm 2.2.

Algorithm 2.2 Pruning Algorithm Mapping $\mathbb{L}_k \to \hat{\mathbb{L}}_k$ Based on the S-Procedure

Input: $\boldsymbol{P}_l$, $l \in \mathbb{L}_k$ // *DRE solutions before pruning, possibly sorted by trace*
Output: $\boldsymbol{P}_{\hat{l}}$, $\hat{l} \in \hat{\mathbb{L}}_k$ // *DRE solutions after pruning*
  $\hat{\mathbb{L}}_k = \{1\}$ // *initialize pruned controller index set* $\hat{\mathbb{L}}_k$
  for $m = 2$ to L_k **do**
    if $\nexists\, \xi_{\hat{l}} \geq 0, \hat{l} \in \hat{\mathbb{L}}_k$: $\sum_{\hat{l}} \xi_{\hat{l}} = 1$, $\boldsymbol{P}_m \succeq \sum_{\hat{l}} \xi_{\hat{l}} \boldsymbol{P}_{\hat{l}}$ **then**
      $\hat{\mathbb{L}}_k = \hat{\mathbb{L}}_k \cup \{m\}$ // *retain controller index*
    end if
  end for

The number of DRE solutions is reduced to $|\hat{\mathbb{L}}_k| \leq |\mathbb{L}_k|$ when pruning is applied at time instant k. By iterating the DRE (2.3) backwards, the number of DRE solutions is given by $|\mathbb{L}_{k-1}| = M|\hat{\mathbb{L}}_k|$. Pruning can then be performed again at time instant $k-1$, leading to $|\hat{\mathbb{L}}_{k-1}| \leq |\mathbb{L}_{k-1}|$. In this way the complexity for solving Problem 2.1 can be reduced considerably.

In the following section a method to further reduce the complexity by relaxing the optimality is presented. The method is based on relaxed dynamic programming which was introduced in [LR06] and [Ran06] for general dynamic programming problems.

2.4 Relaxed Dynamic Programming Solution

For relaxing the optimality, Problem 2.1 is generalized to

Problem 2.7 *For the switched system (1.3) find a control sequence $\boldsymbol{u}(k)$ and a switching sequence $j(k)$ with $k = 0, \ldots, N-1$ such that the resulting cost J_N over the time horizon N is bounded by*

$$J_N^* \leq J_N \leq \alpha J_N^* \tag{2.32}$$

where $J_N^* = \min_{l \in \mathbb{L}_0} \boldsymbol{x}_0^T \boldsymbol{P}_l \boldsymbol{x}_0$ is the minimum cost related to Problem 2.1 and $\alpha \in \mathbb{R}$ with $\alpha \geq 1$ is a relaxation factor.

Problem 2.7 can be approached by relaxed dynamic programming. The main idea of relaxed dynamic programming consists in replacing the Bellman equation (2.11) by the relaxed Bellman equation

$$\underline{V}_{N-k}(k) \leq V_{N-k}(k) \leq \overline{V}_{N-k}(k) \tag{2.33}$$

with the lower bound

$$\underline{V}_{N-k}(k) = \min_{\substack{\boldsymbol{u}(k)\\ j(k)}} \left\{ \ell(k) + V_{N-(k+1)}(k+1) \right\}, \tag{2.34}$$

the upper bound

$$\overline{V}_{N-k}(k) = \min_{\substack{\boldsymbol{u}(k)\\ j(k)}} \left\{ \alpha\ell(k) + V_{N-(k+1)}(k+1) \right\} \tag{2.35}$$

and the terminal condition

$$V_0(N) = \boldsymbol{x}^T(N)\boldsymbol{Q}_0\boldsymbol{x}(N). \tag{2.36}$$

The upper bound contains the relaxation factor α. By introducing inequalities instead of equalities, optimality is relaxed but complexity is reduced since the relaxed value function $V_{N-k}(k)$ can usually be parametrized by a smaller number of quadratic forms than the value function $V_{N-k}^*(k)$. The relaxed value function satisfies

$$V_{N-k}^*(k) \leq \underline{V}_{N-k}(k) \leq V_{N-k}(k) \leq \overline{V}_{N-k}(k) \leq \alpha V_{N-k}^*(k). \tag{2.37}$$

Thus, the gap between the suboptimal value and the optimal value is bounded and characterized by the relaxation factor α.

The objective now is to find a relaxed value function such that the relaxed Bellman equation (2.33) is fulfilled. Considering Problem 2.7, an appropriate parametrization of the relaxed value function is

$$V_{N-k}(k) = \min_{\hat{l} \in \hat{\mathbb{L}}_k} \boldsymbol{x}^T(k)\boldsymbol{P}_{\hat{l}}\boldsymbol{x}(k) \tag{2.38}$$

with $\hat{l} \in \hat{\mathbb{L}}_k \subseteq \mathbb{L}_k$ and $|\hat{\mathbb{L}}_k|$ as small as possible. To determine the matrices $\boldsymbol{P}_{\hat{l}}$, $\hat{l} \in \hat{\mathbb{L}}_k$, the lower and upper bound of the relaxed value function are used. From Theorem 2.2 these are computed as

$$\underline{V}_{N-k}(k) = \min_{l \in \mathbb{L}_k} \boldsymbol{x}^T(k)\underline{\boldsymbol{P}}_l\boldsymbol{x}(k) \tag{2.39a}$$
$$\overline{V}_{N-k}(k) = \min_{l \in \mathbb{L}_k} \boldsymbol{x}^T(k)\overline{\boldsymbol{P}}_l\boldsymbol{x}(k) \tag{2.39b}$$

where $\underline{\boldsymbol{P}}_{l(k)}$ and $\overline{\boldsymbol{P}}_{l(k)}$ with $l(k) \in \mathbb{L}_k$ are solutions to the DREs

$$\begin{aligned}\underline{\boldsymbol{P}}_{l(k)} =&(\boldsymbol{A}_{j(k)} - \boldsymbol{B}_{j(k)}\underline{\boldsymbol{K}}_{l(k)})^T\boldsymbol{P}_{\hat{l}(k+1)}(\boldsymbol{A}_{j(k)} - \boldsymbol{B}_{j(k)}\underline{\boldsymbol{K}}_{l(k)})\\ &+ \boldsymbol{Q}_{1j(k)} + \underline{\boldsymbol{K}}_{l(k)}^T\boldsymbol{Q}_{2j(k)}\underline{\boldsymbol{K}}_{l(k)}\end{aligned} \tag{2.40a}$$
$$\begin{aligned}\overline{\boldsymbol{P}}_{l(k)} =&(\boldsymbol{A}_{j(k)} - \boldsymbol{B}_{j(k)}\overline{\boldsymbol{K}}_{l(k)})^T\boldsymbol{P}_{\hat{l}(k+1)}(\boldsymbol{A}_{j(k)} - \boldsymbol{B}_{j(k)}\overline{\boldsymbol{K}}_{l(k)})\\ &+ \alpha\boldsymbol{Q}_{1j(k)} + \alpha\overline{\boldsymbol{K}}_{l(k)}^T\boldsymbol{Q}_{2j(k)}\overline{\boldsymbol{K}}_{l(k)}\end{aligned} \tag{2.40b}$$

with

$$\boldsymbol{P}_{\hat{l}(N)} = \boldsymbol{Q}_0 \tag{2.41a}$$
$$\underline{\boldsymbol{K}}_{l(k)} = (\boldsymbol{Q}_{2j(k)} + \boldsymbol{B}_{j(k)}^T\boldsymbol{P}_{\hat{l}(k+1)}\boldsymbol{B}_{j(k)})^{-1}\boldsymbol{B}_{j(k)}^T\boldsymbol{P}_{\hat{l}(k+1)}\boldsymbol{A}_{j(k)} \tag{2.41b}$$
$$\overline{\boldsymbol{K}}_{l(k)} = (\alpha\boldsymbol{Q}_{2j(k)} + \boldsymbol{B}_{j(k)}^T\boldsymbol{P}_{\hat{l}(k+1)}\boldsymbol{B}_{j(k)})^{-1}\boldsymbol{B}_{j(k)}^T\boldsymbol{P}_{\hat{l}(k+1)}\boldsymbol{A}_{j(k)}. \tag{2.41c}$$

A simple approach to obtain matrices $\boldsymbol{P}_{\hat{l}}$ such that the relaxed Bellman equation (2.33) is satisfied then consists in using some of the lower bound DRE solutions $\underline{\boldsymbol{P}}_l$ as the matrices $\boldsymbol{P}_{\hat{l}}$, leading to the following relaxation criterion:

Theorem 2.8 (Relaxation Criterion) *If for each* $\overline{\boldsymbol{P}}_m$ *with* $m \in \mathbb{L}_k$ *and each* $\boldsymbol{x}(k) \in \mathbb{R}^n$ *there exists at least one* $\boldsymbol{P}_{\hat{l}} \in \{\underline{\boldsymbol{P}}_l, l \in \mathbb{L}_k\}$ *with* $\hat{l} \in \hat{\mathbb{L}}_k \subseteq \mathbb{L}_k$ *such that*

$$\boldsymbol{x}^T(k)\overline{\boldsymbol{P}}_m\boldsymbol{x}(k) \geq \boldsymbol{x}^T(k)\boldsymbol{P}_{\hat{l}}\boldsymbol{x}(k), \tag{2.42}$$

then the relaxed value function (2.38) satisfies the relaxed Bellman equation (2.33).

PROOF. By construction. A schematic illustration of the relaxation criterion is given in Figure 2.6. The lower/upper bound $\underline{V}_{N-k}(\boldsymbol{x}(k))/\overline{V}_{N-k}(\boldsymbol{x}(k))$ of the relaxed value function is plotted black/gray, bold and dashed. The constituting quadratic forms $\boldsymbol{x}^T(k)\underline{\boldsymbol{P}}_l\boldsymbol{x}(k)$ / $\boldsymbol{x}^T(k)\overline{\boldsymbol{P}}_l\boldsymbol{x}(k)$ with $l \in \{1, 2, 3\}$ are depicted black/gray, non-bold and dashed, dotted or dash-dotted. The relaxed value function $V_{N-k}(\boldsymbol{x}(k))$ satisfying the relaxed Bellman equation (2.33) can obviously be constructed from the quadratic forms $\boldsymbol{x}^T(k)\underline{\boldsymbol{P}}_l\boldsymbol{x}(k)$ with $l \in \{1, 3\}$ since, loosely speaking, the relaxed value function then runs between the lower and upper bound. Consequently, the quadratic form $\boldsymbol{x}^T(k)\underline{\boldsymbol{P}}_l\boldsymbol{x}(k)$ with $l = 2$ is dispensable. The relaxed value function is plotted light gray, bold and solid. Note that this relaxed value function fulfills the relaxed Bellman equation also for smaller relaxation factors α, precisely until the upper bound falls below the point indicated by the arrow. □

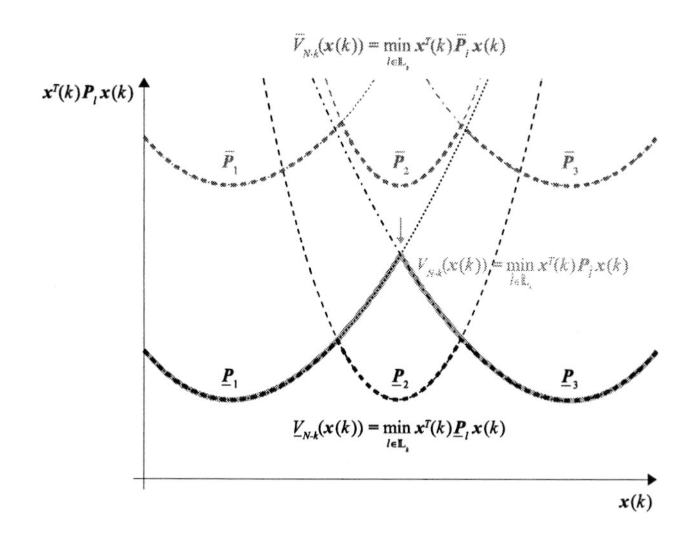

Figure 2.6: Illustration of the Relaxation Criterion

Remark 2.9. The relaxation criterion can not be evaluated exactly since the controller index $\hat{l} \in \hat{\mathbb{L}}_k \setminus \{m\}$ fulfilling the condition (2.42) may vary over $\boldsymbol{x}(k) \in \mathbb{R}^n$. Instead, a common controller index $n \in \hat{\mathbb{L}}_k \setminus \{m\}$ can be considered for all states $\boldsymbol{x}(k) \in \mathbb{R}^n$ analogously to Theorem 2.5. Alternatively, a convex combination of $\boldsymbol{P}_{\hat{l}}$ can be utilized deploying the S-procedure analogously to Theorem 2.6. The latter approach is less conservative but computationally more demanding as pointed out in Remarks 2.7 and 2.8. For brevity of presentation only the latter approach will be considered in the following.

Remark 2.10. The upper bound DRE solution $\overline{\boldsymbol{P}}_l$ is always dominated by the associated lower bound DRE solution $\underline{\boldsymbol{P}}_l$, i.e. $\underline{\boldsymbol{P}}_l \preceq \overline{\boldsymbol{P}}_l$. Therefore, (2.42) is satisfied trivially for $\hat{\mathbb{L}}_k = \mathbb{L}_k$. In terms of complexity reduction it is rather desired to fulfill (2.42) with a minimum number of DRE solutions $|\hat{\mathbb{L}}_k|$.

An efficient domination criterion based on the S-procedure can be formulated as

Theorem 2.9 (Domination Criterion) *If there exist scalars $\xi_{\hat{l}} \geq 0$ and $\boldsymbol{P}_{\hat{l}} \in \{\underline{\boldsymbol{P}}_l, l \in \mathbb{L}_k\}$ with $\hat{l} \in \hat{\mathbb{L}}_k$ such that*

$$\sum_{\hat{l} \in \hat{\mathbb{L}}_k} \xi_{\hat{l}} = 1 \quad \textit{and} \quad \overline{\boldsymbol{P}}_m \succeq \sum_{\hat{l} \in \hat{\mathbb{L}}_k} \xi_{\hat{l}} \boldsymbol{P}_{\hat{l}}, \tag{2.43}$$

then $\overline{\boldsymbol{P}}_m$ is dominated.

Proof. Follows the same lines as the proof of Theorem 2.6. □

An algorithm for constructing the relaxed value function $V_{N-k}(k)$ using Theorem 2.9 is given in Algorithm 2.3.

Algorithm 2.3 Relaxed Pruning Algorithm Based on the S-Procedure

Input: $\underline{\boldsymbol{P}}_l, \overline{\boldsymbol{P}}_l, l \in \mathbb{L}_k$ *// lower/upper bound DRE solutions, possibly sorted as in (2.44)*
Output: $\boldsymbol{P}_{\hat{l}}, \hat{l} \in \hat{\mathbb{L}}_k$ *// DRE solutions after relaxed pruning*

$\hat{\mathbb{L}}_k = \{1\}$ *// initialize pruned controller index set* $\hat{\mathbb{L}}_k$
for $m = 2$ to L_k **do**
  if $\nexists\, \xi_{\hat{l}} \geq 0, \hat{l} \in \hat{\mathbb{L}}_k$: $\sum\limits_{\hat{l}} \xi_{\hat{l}} = 1, \overline{\boldsymbol{P}}_m \succeq \sum\limits_{\hat{l}} \xi_{\hat{l}} \boldsymbol{P}_{\hat{l}}$ **then**
    $\hat{\mathbb{L}}_k = \hat{\mathbb{L}}_k \cup \{m\}$ *// add controller index*
    $\boldsymbol{P}_m = \underline{\boldsymbol{P}}_m$ *// add lower bound DRE solution*
  end if
end for

The algorithm is based on adding lower bound DRE solutions $\underline{\boldsymbol{P}}_l$, $l \in \mathbb{L}_k$ until all upper bound DRE solutions $\overline{\boldsymbol{P}}_l$, $l \in \mathbb{L}_k$ are dominated. Therefor the condition (2.43) is evaluated. It is again essential to initially add DRE solutions which are presumably not dominated. This can be achieved by presorting the lower bound DRE solutions by trace, i.e. reassigning the controller indices $l \in \mathbb{L}_k$ of $\underline{\boldsymbol{P}}_l$, $\overline{\boldsymbol{P}}_l$ such that

$$\mathrm{tr}(\underline{\boldsymbol{P}}_1) \leq \mathrm{tr}(\underline{\boldsymbol{P}}_2) \leq \ldots \leq \mathrm{tr}(\underline{\boldsymbol{P}}_{L_k}). \tag{2.44}$$

Remark 2.11. The choice of α is a compromise between suboptimality and complexity. If α is close to 1, then the relaxed Bellman equation (2.33) approaches the Bellman equation (2.11), resulting in a close-to-optimal solution. However, the relaxed value function $V_{N-k}(k)$ will be only marginally simpler than the value function $V^*_{N-k}(k)$, leading to a low complexity reduction. If α is considerably larger than 1, the converse holds. Note that for $\alpha = 1$ the Relaxed Pruning Algorithm 2.3 is equivalent to the Pruning Algorithm 2.2 introduced in Section 2.3.

Remark 2.12. The relaxation factor is contained in (2.35) in a multiplicative manner. The relaxation factor can also be introduced in an additive manner, i.e.

$$\overline{V}_{N-k}(k) = \min_{\substack{\boldsymbol{u}(k) \\ j(k)}} \left\{ \ell(k) + \alpha + V_{N-(k+1)}(k+1) \right\} \tag{2.45}$$

with $\alpha \geq 0$ as discussed in [LR02]. This approach, however, has some drawbacks: The absolute suboptimality must be predefined instead of the relative suboptimality which is difficult since the magnitude of the relaxed value function is not known in advance. Furthermore, the suboptimality increases while iterating the relaxed value function backwards.

Remark 2.13. Besides the parametrization of the relaxed value function as in (2.38) also alternative parametrizations can be sought to improve the effectiveness of relaxed dynamic programming, e.g. based on positive polynomials as proposed in [Wer07].

In the following the properties and effectiveness of relaxed dynamic programming are studied by a numerical example and a numerical study.

Example 2.2 Reconsider the setup of Example 2.1 for $\boldsymbol{Q}_{21} = \boldsymbol{Q}_{22} = \boldsymbol{Q}_{23} = \boldsymbol{Q}_{24} = 1$. Problem 2.7 is solved via relaxed dynamic programming using Algorithm 2.3. The number of DRE solutions with relaxed pruning $|\hat{\mathbb{L}}_k|$ and without relaxed pruning $|\mathbb{L}_k|$ as a function of the steps-to-go $N-k$ and the relaxation factor α is shown in Figure 2.7.

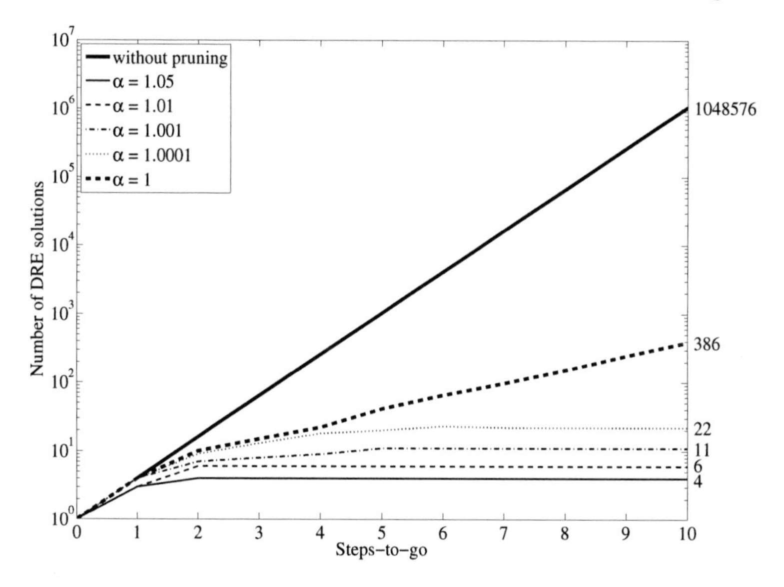

Figure 2.7: Number of DRE solutions

Obviously, Problem 2.7 is computationally intractable if no relaxed pruning is applied since $|\mathbb{L}_0| = 1048576$. For $\alpha = 1$, the number of DRE solutions extremely reduces to $|\hat{\mathbb{L}}_0| = 386$. For $\alpha = 1.0001$, the number of DRE solutions further reduces to $|\hat{\mathbb{L}}_0| = 22$ and decreases even more with increasing α. Complexity can thus be reduced substantially while relaxing optimality only marginally. Noteworthy, when relaxed pruning is applied the number of DRE solutions converges after several optimization steps. The larger α, the less optimization steps are required for convergence.

The value function $V_{10}^*(\boldsymbol{x}(0))$, the relaxed value function $V_{10}(\boldsymbol{x}(0))$ and the upscaled value function $\alpha V_{10}^*(\boldsymbol{x}(0))$ for $\alpha = 1.05$ are plotted in Figure 2.8 on the unit semi-circle $\boldsymbol{x}(0) = \begin{pmatrix}\cos\varphi & \sin\varphi\end{pmatrix}^T$ with $\varphi \in [0, \pi]$. It can be observed that inequality (2.37) is fulfilled. Furthermore, the relaxed value function $V_{10}(\boldsymbol{x}(0))$ touches both the lower bound $V_{10}^*(\boldsymbol{x}(0))$ and the upper bound $\alpha V_{10}^*(\boldsymbol{x}(0))$ of inequality (2.37). This implies that the sufficient domination criterion proposed in Theorem 2.9 is not overly conservative.

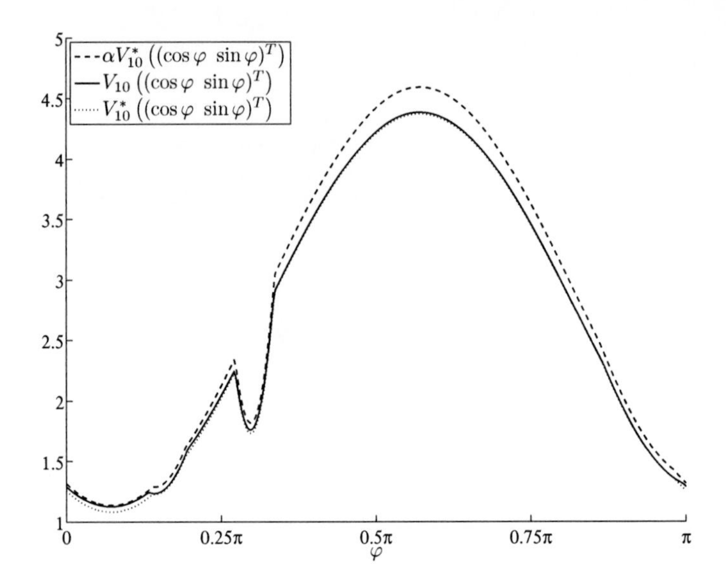

Figure 2.8: Value function, relaxed value function and upscaled value function

□

Example 2.3 Hundred switched systems (1.3) are generated for each $M \in \{2,\ldots,6\}$, $\mathrm{n} \in \{2,3\}$, $\mathrm{m} = 1$ and each $M = 2$, $\mathrm{n} \in \{4,\ldots,6\}$, $\mathrm{m} = 1$ which have stabilizable subsystems with random eigenvalues $\lambda_{1,\ldots,\mathrm{n}}$ fulfilling $\lambda_{1,\ldots,\mathrm{n}} \in \mathbb{C}$ with $|\lambda_{1,\ldots,\mathrm{n}}| \leq 1.25$ and $\lambda_{1,\ldots,\mathrm{n}} \notin \mathbb{R}^-$. Hence, some subsystems are unstable. For each switched system, Problem 2.7 is solved for $\boldsymbol{Q}_0 = \boldsymbol{0}$, $\boldsymbol{Q}_{1j(k)} = \boldsymbol{I}$, $\boldsymbol{Q}_{2j(k)} = 1\ \forall j(k) \in \mathbb{M}$, $N = 6$ and $\alpha = 1.02$ by relaxed dynamic programming utilizing Algorithm 2.3.

The absolute frequency of the number of DRE solutions $|\hat{\mathbb{L}}_0|$ obtained from relaxed dynamic programming is plotted in Figures 2.9, 2.10 and 2.11 on pages 37 and 38. The median (indicated by $\triangle$) of the absolute frequency grows proportionally with the number of subsystems M as can be seen in Figures 2.9 and 2.10. This suggests that the polynomial complexity of Problem 2.7 with respect to M can be practically reduced to a linear complexity by relaxed dynamic programming. Furthermore, the median of the absolute frequency increases overproportionally with the system order n as can be observed from Figure 2.11 which implies that the effectiveness of relaxed dynamic programming reduces with increasing order. The interquartile range between the first and the third quartile (indicated by $\triangleright$ and $\triangleleft$) generally becomes wider both for increasing M and n. Overall, the number of DRE solutions remains acceptable even for large M and n. The computation times are indicated in Example 3.6 on page 57. □

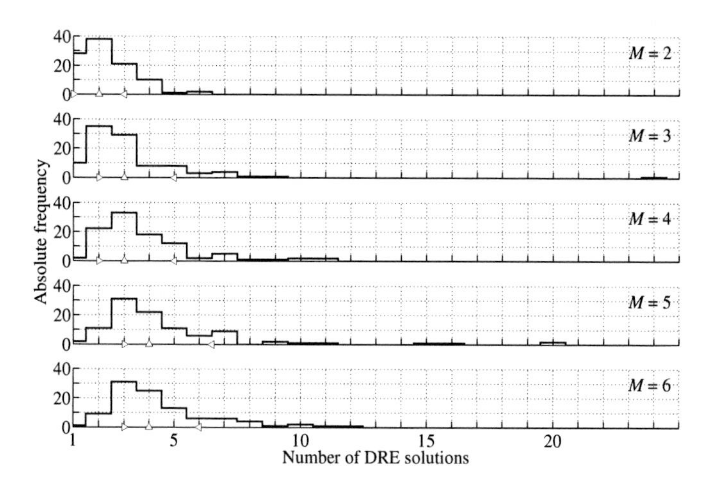

Figure 2.9: Absolute frequency of the number of DRE solutions for n = 2

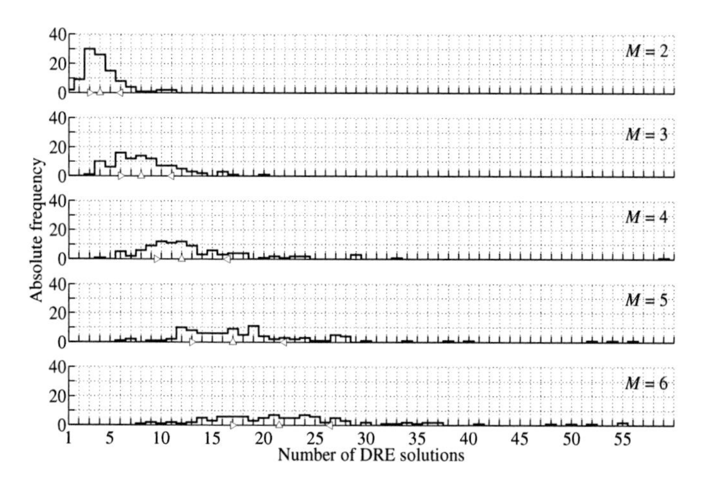

Figure 2.10: Absolute frequency of the number of DRE solutions for n = 3

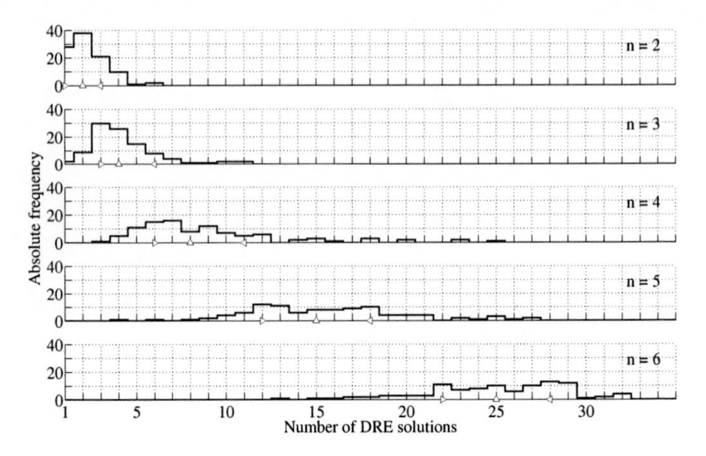

Figure 2.11: Absolute frequency of the number of DRE solutions for M = 2

In the following chapter a receding-horizon control and scheduling strategy is formulated based on the finite-horizon solution.

3 Receding-Horizon Control and Scheduling

3.1 Problem Formulation

Consider the quadratic discrete-time cost function

$$\begin{aligned} J_N(k) =& \boldsymbol{x}^T(k+N)\boldsymbol{Q}_0\boldsymbol{x}(k+N) \\ &+ \sum_{i=0}^{N-1} \left[\boldsymbol{x}^T(k+i)\boldsymbol{Q}_{1j(k+i)}\boldsymbol{x}(k+i) + \boldsymbol{u}^T(k+i)\boldsymbol{Q}_{2j(k+i)}\boldsymbol{u}(k+i)\right] \end{aligned} \tag{3.1}$$

where $\boldsymbol{Q}_0 \succeq \boldsymbol{0}$ is a symmetric terminal weighting matrix, $\boldsymbol{Q}_{1j(k)} \succeq \boldsymbol{0}$ and $\boldsymbol{Q}_{2j(k)} \succ \boldsymbol{0}$ with $j(k) \in \mathbb{M}$ are symmetric switched weighting matrices and N is the prediction horizon. Furthermore, k denotes the current time instant, i.e. the absolute time, and i denotes the time instant within the prediction horizon, i.e. the relative time, cf. Section 1.4.4.

Problem 3.1 *For the switched system (1.3) and the current state $\boldsymbol{x}(k)$ find a control sequence $\boldsymbol{u}^*(k), \ldots, \boldsymbol{u}^*(k+N-1)$ and a switching sequence $j^*(k), \ldots, j^*(k+N-1)$ such that the cost function (3.1) is minimized over the prediction horizon N, i.e.*

$$\begin{aligned} &\min_{\substack{\boldsymbol{u}(k),\ldots,\boldsymbol{u}(k+N-1) \\ j(k),\ldots,j(k+N-1)}} J_N(k) \\ &\text{subject to } \boldsymbol{x}(k+1+i) = \boldsymbol{A}_{j(k+i)}\boldsymbol{x}(k+i) + \boldsymbol{B}_{j(k+i)}\boldsymbol{u}(k+i),\ i = 0, \ldots, N-1. \end{aligned} \tag{3.2}$$

If the first element $\boldsymbol{u}^*(k)$ of the control sequence and the first element $j^*(k)$ of the switching sequence are then applied to the switched system at the current time instant k, a receding-horizon control and scheduling (RHCS) strategy is obtained.

Remark 3.1. Optimality can be relaxed analogously to Problem 2.7 which leads to a relaxed receding-horizon control and scheduling (RRHCS) strategy.

Remark 3.2. Receding-horizon control and scheduling may be seen as an approximation of infinite-horizon control and scheduling as outlined in Section 1.4.4. Quantifying the suboptimality induced by this approximation, specifically the relation between the finite-horizon cost $J_N(k)$ according to (3.1) and the infinite-horizon cost J_∞ according to (2.1), is surely desirable. This issue has rarely been addressed in the literature. Notable exceptions are [NP97, PN00, GR08, Grü09] where suboptimality criteria for receding-horizon control of linear and nonlinear systems have been formulated. These criteria may be revised for receding-horizon control and scheduling of switched systems. Specifically the suboptimality induced by a relaxation may also be regarded in such criteria. This is, however, out of the scope of this thesis.

3.2 Explicit Solution

Problem 3.1 can be transformed into Problem 2.1 by using the current state $\boldsymbol{x}(k)$ as the initial state $\boldsymbol{x}_0$. The solution then follows from Theorem 2.2, where only the first element of the DRE solution $\boldsymbol{P}_{l(0)}$ with $l(0) \in \mathbb{L}_0$, the control sequence $\boldsymbol{u}^*(0) = -\boldsymbol{K}_{l^*(0)}\boldsymbol{x}(k)$ and the switching sequence $j^*(0) = \sigma(l^*(0))$ respectively is used in the RHCS strategy. The indication of the first element of the controller sequence $l(0)$ and of the switching sequence $j(0)$ is omitted in the following for notational convenience.

The RHCS strategy can then be reformulated as follows:

Theorem 3.2 *The solution to Problem 3.1 is given by the piecewise linear (PWL) state feedback control law*

$$\boldsymbol{u}^*(k) = -\boldsymbol{K}_m\boldsymbol{x}(k) \quad \text{for} \quad \boldsymbol{x}(k) \in \mathcal{X}_m \tag{3.3}$$

where the regions $\mathcal{X}_m$ with $\bigcup_{l\in\mathbb{L}_0} \mathcal{X}_l = \mathbb{R}^n$ are described by

$$\mathcal{X}_m = \left\{\boldsymbol{x}(k) \mid \boldsymbol{x}^T(k)\boldsymbol{P}_m\boldsymbol{x}(k) \leq \boldsymbol{x}^T(k)\boldsymbol{P}_l\boldsymbol{x}(k)\ \forall l; m,l \in \mathbb{L}_0\right\}. \tag{3.4}$$

PROOF. According to Theorem 2.2 and in particular (2.5) the optimal controller index results for the current state $\boldsymbol{x}(k)$ as

$$m = l^* = \arg\min_{l\in\mathbb{L}_0} \boldsymbol{x}^T(k)\boldsymbol{P}_l\boldsymbol{x}(k). \tag{3.5}$$

This corresponds to

$$\boldsymbol{x}^T(k)\boldsymbol{P}_m\boldsymbol{x}(k) \leq \boldsymbol{x}^T(k)\boldsymbol{P}_l\boldsymbol{x}(k) \quad \forall l \in \mathbb{L}_0 \tag{3.6}$$

which leads to the regions (3.4). □

Remark 3.3. The PWL state feedback control law (3.3) is *explicit*. The feedback matrices $\boldsymbol{K}_l$ and the DRE solutions $\boldsymbol{P}_l$ can be calculated offline for all $l \in \mathbb{L}_0$ using any of the methods introduced in Chapter 2 and stored. The online computation reduces to the evaluation of the PWL state feedback control law (3.3) which includes a region membership test based on (3.4) or (3.5).

Remark 3.4. Two regions $\mathcal{X}_m$ and $\mathcal{X}_n$ with $m, n \in \mathbb{L}_0$ only overlap on their boundary

$$\partial\mathcal{X}_{mn} = \left\{\boldsymbol{x}(k) \mid \boldsymbol{x}^T(k)\boldsymbol{P}_m\boldsymbol{x}(k) = \boldsymbol{x}^T(k)\boldsymbol{P}_n\boldsymbol{x}(k) \leq \boldsymbol{x}^T(k)\boldsymbol{P}_l\boldsymbol{x}(k)\ \forall l \in \mathbb{L}_0 \setminus \{m,n\}\right\}. \tag{3.7}$$

The value function, however, takes the same value for both controller indices m and n on the boundary $\partial\mathcal{X}_{mn}$; hence, m or n may be chosen arbitrarily. The actual choice depends on the implementation of the region membership test. The explicit solution (3.3) is therefore well-posed.

Remark 3.5. The regions $\mathcal{X}_l$ with $l \in \mathbb{L}_0$ are usually non-convex. Exceptions are e.g. second-order switched systems where the regions are piecewise convex, i.e. composed of convex subregions where the boundaries of the subregions are given by lines through the origin as shown in Appendix A.3.

Remark 3.6. Substituting (3.3) into (1.3) leads to the autonomous PWL closed-loop system

$$\boldsymbol{x}(k+1) = \widetilde{\boldsymbol{A}}_m \boldsymbol{x}(k) \quad \text{for} \quad \boldsymbol{x}(k) \in \mathcal{X}_m, \quad \boldsymbol{x}(0) = \boldsymbol{x}_0 \tag{3.8}$$

with $\widetilde{\boldsymbol{A}}_m = \boldsymbol{A}_v - \boldsymbol{B}_v \boldsymbol{K}_m$ where the switching index $v \in \mathbb{M}$ follows from $v = \sigma(m)$ for a known region $\mathcal{X}_m$ with $m \in \mathbb{L}_0$.

Remark 3.7. The explicit PWL state feedback control law (3.3) and PWL closed-loop system (3.8) can be formulated analogously with respect to the RRHCS strategy using $\hat{l} \in \hat{\mathbb{L}}_0$ instead of $l \in \mathbb{L}_0$ where $\hat{l}$ indexes the DRE solutions and switched feedback matrices after relaxed pruning.

Example 3.1 Reconsider the setup of Example 2.2 in terms of the RRHCS strategy (3.3) with relaxation factor $\alpha = 1.05$. The regions $\mathcal{X}_m$ with $m \in \hat{\mathbb{L}}_0$ and $|\hat{\mathbb{L}}_0| = 4$, which are piecewise convex for second-order switched systems as outlined in Remark 3.5, are shown in Figure 3.1. Moreover, the trajectories of the PWL closed-loop system (3.8) are plotted for different initial states.

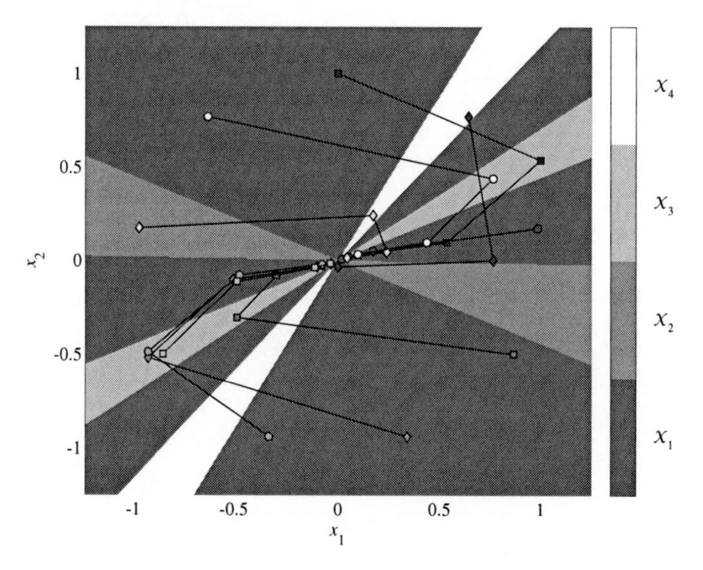

Figure 3.1: Regions and closed-loop trajectories for different initial states

The switching function is given by

$$j = \sigma(\hat{l}) = \begin{cases} 2 & \text{if } \hat{l} = 3 \\ 3 & \text{if } \hat{l} = 1, 2 \\ 4 & \text{if } \hat{l} = 4, \end{cases} \tag{3.9}$$

hence subsystem with the switching index $j = 1$ will never be active. E.g. for the initial state $\boldsymbol{x}_0 = \begin{pmatrix} 0 & 1 \end{pmatrix}^T$ the controller and switching sequence

$$\hat{l}(k) = (1, 3, 1, 1, 1, \ldots) \tag{3.10a}$$
$$j(k) = (3, 2, 3, 3, 3, \ldots) \tag{3.10b}$$

result respectively. Evidently, neither all controller indices $\hat{l} \in \hat{\mathbb{L}}_0$ nor all switching indices $j \in \mathbb{M}$ occur therein. □

3.3 Stability Analysis

Global uniform asymptotic stability of the switched system (1.3) under the (R)RHCS strategy (3.3) is not guaranteed inherently. Two different approaches to handle the stability of the PWL closed-loop system (3.8) are introduced in this section. First, an *a priori* stability condition is derived which applies to the RHCS strategy only. Here, the stability of the PWL closed-loop system is ensured before the RHCS strategy is determined. Second, several *a posteriori* stability criteria for the already determined PWL closed-loop system are devised which apply to both the RHCS and the RRHCS strategy. The criteria are based on constructing a piecewise quadratic (PWQ) Lyapunov function or on utilizing the value function as a candidate Lyapunov function.

3.3.1 A Priori Stability Condition

Closed-Loop Stability under the RHCS Strategy

Since the terminal weighting matrix $\boldsymbol{Q}_0$ can be chosen arbitrarily in the quadratic cost function (3.1), it can be used to influence the RHCS strategy in a way that the obtained PWL closed-loop system (3.8) is globally uniformly asymptotically stable for an arbitrary time horizon N, refer also to [MR03, GKBM05, LHWB06, Laz06].

Theorem 3.3 *Consider the suboptimal state feedback control law*

$$\boldsymbol{u}(k) = -\boldsymbol{K}_v \boldsymbol{x}(k), \quad v \in \mathbb{M}. \tag{3.11}$$

The switched system (1.3) *under the RHCS strategy* (3.3) *is globally uniformly asymptotically stable if the switched weighting matrices $\boldsymbol{Q}_{1j(k)}$ are strictly positive definite for* *all $j(k) \in \mathbb{M}$ and there exist a terminal weighting matrix $\boldsymbol{Q}_0 \succ \boldsymbol{0}$ and switched feedback matrices $\boldsymbol{K}_v$ satisfying*

$$\boldsymbol{Q}_0 - (\boldsymbol{A}_v - \boldsymbol{B}_v\boldsymbol{K}_v)^T\boldsymbol{Q}_0(\boldsymbol{A}_v - \boldsymbol{B}_v\boldsymbol{K}_v) - \boldsymbol{Q}_{1v} - \boldsymbol{K}_v^T\boldsymbol{Q}_{2v}\boldsymbol{K}_v \succ \boldsymbol{0} \quad \forall v \in \mathbb{M}. \tag{3.12}$$

Proof. Introduce the suboptimal value function $V_N(k) = V_N(\boldsymbol{x}(k))$ described by

$$V_N(k) = \boldsymbol{x}^T(k+N)\boldsymbol{Q}_0\boldsymbol{x}(k+N) + \sum_{i=0}^{N-1} \ell(k+i) \tag{3.13}$$

with the non-negative step cost $\ell(k+i)$ as defined in (2.10). The suboptimal value function defines the cost for some switching sequence $j(k), \ldots, j(k+N-1)$ and some control sequence $\boldsymbol{u}(k), \ldots, \boldsymbol{u}(k+N-1)$ between the time instants k and $k+N$. The value function

$$J_N^*(k) = V_N^*(k) = \min_{\substack{\boldsymbol{u}(k),\ldots,\boldsymbol{u}(k+N-1) \\ j(k),\ldots,j(k+N-1)}} V_N(k) \tag{3.14}$$

is obtained as solution to Problem 3.1.

For ensuring global uniform asymptotic stability, require that the value function $V_N^*(k)$ is a Lyapunov function for the PWL closed-loop system (3.8). The value function $V_N^*(k)$ is positive definite, decrescent and radially unbounded due to requiring $\boldsymbol{Q}_{1j(k)}$ strictly positive definite for all $j(k) \in \mathbb{M}$. The difference of the value function $V_N^*(k)$ can be made negative definite by demanding

$$V_N^*(k+1) - V_N^*(k) \overset{!}{<} -\ell^*(k) < 0 \quad \forall \boldsymbol{x}(k) \in \mathbb{R}^n \setminus \{\boldsymbol{0}\} \quad \forall k \in \mathbb{N}_0. \tag{3.15}$$

To check the condition (3.15), Problem 3.1 has not to be solved. Rather, the suboptimal state feedback control law

$$\boldsymbol{u}(k+N) = -\boldsymbol{K}_{j(k+N)}\boldsymbol{x}^*(k+N) \tag{3.16}$$

can be used at the final predicted time instant $k+N$ with $j(k+N) \in \mathbb{M}$, leading to the suboptimal value function $V_N(k+1) \geq V_N^*(k+1)$. Consequently, if the inequality

$$V_N(k+1) - V_N^*(k) \overset{!}{<} -\ell^*(k) \quad \forall \boldsymbol{x}(k) \in \mathbb{R}^n \setminus \{\boldsymbol{0}\} \quad \forall k \in \mathbb{N}_0 \tag{3.17}$$

holds, then the inequality (3.15) is also fulfilled. Substituting (3.13) and (3.14) into (3.17) and canceling out the sums yields

$$\begin{aligned} V_N(k+1) - V_N^*(k) = {} & \boldsymbol{x}^T(k+N+1)\boldsymbol{Q}_0\boldsymbol{x}(k+N+1) \\ & + \boldsymbol{u}^T(k+N)\boldsymbol{Q}_{2j(k+N)}\boldsymbol{u}(k+N) \\ & - \boldsymbol{x}^{*^T}(k+N)\left(\boldsymbol{Q}_0 - \boldsymbol{Q}_{1j(k+N)}\right)\boldsymbol{x}^*(k+N) \\ & - \boldsymbol{x}^{*^T}(k)\boldsymbol{Q}_{1j^*(k)}\boldsymbol{x}^*(k) - \boldsymbol{u}^{*^T}(k)\boldsymbol{Q}_{2j^*(k)}\boldsymbol{u}^*(k) \\ \overset{!}{<} {} & -\ell^*(k) \quad \forall \boldsymbol{x}(k) \in \mathbb{R}^n \setminus \{\boldsymbol{0}\} \quad \forall k \in \mathbb{N}_0. \end{aligned} \tag{3.18}$$

Using that $\boldsymbol{x}^{*^T}(k)\boldsymbol{Q}_{1j^*(k)}\boldsymbol{x}^*(k)+\boldsymbol{u}^{*^T}(k)\boldsymbol{Q}_{2j^*(k)}\boldsymbol{u}^*(k)=\ell^*(k)$ and substituting (3.16) and $\boldsymbol{x}(k+N+1)=\boldsymbol{A}_{j(k+N)}\boldsymbol{x}^*(k+N)+\boldsymbol{B}_{j(k+N)}\boldsymbol{u}(k+N)$ into (3.18) yields

$$\begin{aligned} &V_N(k+1)-V_N^*(k)\\ &=\boldsymbol{x}^{*^T}(k+N)\Big((\boldsymbol{A}_v-\boldsymbol{B}_v\boldsymbol{K}_v)^T\boldsymbol{Q}_0\,(\boldsymbol{A}_v-\boldsymbol{B}_v\boldsymbol{K}_v)-\boldsymbol{Q}_0+\boldsymbol{Q}_{1v}+\boldsymbol{K}_v^T\boldsymbol{Q}_{2v}\boldsymbol{K}_v\Big)\boldsymbol{x}^*(k+N)\\ &\overset{!}{<}0\quad\forall\boldsymbol{x}(k)\in\mathbb{R}^n\setminus\{\boldsymbol{0}\}\quad\forall k\in\mathbb{N}_0. \end{aligned}\tag{3.19}$$

with $v=j(k+N)$. The inequality (3.19) holds iff the matrix inequality (3.12) is satisfied. The value function $V_N^*(k)$ is a CQLF for the PWL closed-loop system (3.8) if the matrix inequality (3.12) is satisfied for all switching indices $v\in\mathbb{M}$, thereby proving global uniform asymptotic stability, cf. Section 1.3.2. Problem 3.1 is finally solved with the resulting terminal weighting matrix $\boldsymbol{Q}_0$ using any of the pruning Algorithms 2.1 and 2.2 which preserve optimality. The switched feedback matrices $\boldsymbol{K}_v$ which were introduced for technical reasons only are not used anymore. □

Remark 3.8. The suboptimal state feedback control law (3.11) can also be chosen as $\boldsymbol{u}(k)=\boldsymbol{0}$, simplifying (3.12) to

$$\boldsymbol{Q}_0-\boldsymbol{A}_v^T\boldsymbol{Q}_0\boldsymbol{A}_v-\boldsymbol{Q}_{1v}\succ 0\quad\forall v\in\mathbb{M}.\tag{3.20}$$

Remark 3.9. Instead of a common terminal weighting matrix $\boldsymbol{Q}_0$ also switched terminal weighting matrices $\boldsymbol{Q}_{0j(k+N)}$ (symmetric and positive definite for all $j(k+N)\in\mathbb{M}$) can be considered in the quadratic cost function (3.1) since the switching index $j(k+N)$ is out of the optimization horizon and can therefore be chosen arbitrarily. Following the lines of the proof of Theorem 3.3, it can be shown that the matrix inequality (3.12) in Theorem 3.3 then changes to

$$\boldsymbol{Q}_{0v}-(\boldsymbol{A}_v-\boldsymbol{B}_v\boldsymbol{K}_v)^T\boldsymbol{Q}_{0w}(\boldsymbol{A}_v-\boldsymbol{B}_v\boldsymbol{K}_v)-\boldsymbol{Q}_{1v}-\boldsymbol{K}_v^T\boldsymbol{Q}_{2v}\boldsymbol{K}_v\succ 0\quad\forall v,w\in\mathbb{M}\tag{3.21}$$

with $v=j(k+N)$ and $w=j(k+N+1)$. The value function $V_N^*(k)$ is then an SQLF for the PWL closed-loop system (3.8) if the matrix inequality (3.21) is satisfied for all consecutive switching indices $v,w\in\mathbb{M}$. Obviously, the matrix inequality (3.21) implies the matrix inequality (3.12) when selecting $\boldsymbol{Q}_v=\boldsymbol{Q}_w=\boldsymbol{Q}$. Therefore, the conservatism of Theorem 3.3 can be reduced by using a switched terminal weighting matrix. Problem 3.1 is then solved using dynamic programming with the terminal conditions $\boldsymbol{P}_{l(N)}=\boldsymbol{Q}_{0j(k+N)}$ with $l(N)\in\mathbb{L}_N=\mathbb{M}$.

Remark 3.10. For the existence of a common terminal weighting matrix $\boldsymbol{Q}_0$ or switched terminal weighting matrices $\boldsymbol{Q}_{0v}$ and switched feedback matrices $\boldsymbol{K}_v$ which satisfy (3.12), (3.20) or (3.21), the subsystems $(\boldsymbol{A}_v,\boldsymbol{B}_v)$ of the switched system (1.3) must be stabilizable for all $v\in\mathbb{M}$ (for (3.12) and (3.21), cf. Remark 1.8) or globally asymptotically stable (for (3.20), cf. Section 1.3.2). These preconditions arise due considering arbitrary switching in (3.12), (3.20) and (3.21) and may thus be removed by considering constrained switching, cf. Remark 1.8. Then not only a suboptimal state feedback control law but also a suboptimal state feedback switching law at the final predicted time instant

$k+N$ must be designed. To this end, the a posteriori stability criterion based on the value function given in Theorem 3.9 may be modified for design. Specifically, $\boldsymbol{P}_{\hat{l}}$ and $\boldsymbol{K}_m$ contained in $\tilde{\boldsymbol{A}}_m$ may be considered as design variables. Furthermore, the approach presented in [GC06] may be reconsidered in this regard.

For the sake of generality a switched terminal weighting matrix is considered in the following.

LMI Formulation

To find $\boldsymbol{Q}_{0v}$, $v \in \mathbb{M}$ such that inequality (3.21) is fulfilled, an LMI feasibility problem can be formulated:

Theorem 3.4 *Inequality* (3.21) *is feasible for all $v, w \in \mathbb{M}$ iff there exist symmetric matrices $\boldsymbol{Z}_v \succ \boldsymbol{0}$ and matrices $\boldsymbol{G}_v$ and $\boldsymbol{W}_v$ such that the LMIs*

$$\begin{pmatrix} \boldsymbol{G}_v^T + \boldsymbol{G}_v - \boldsymbol{Z}_v & * & * & * \\ \boldsymbol{A}_v\boldsymbol{G}_v - \boldsymbol{B}_v\boldsymbol{W}_v & \boldsymbol{Z}_w & * & * \\ \boldsymbol{Q}_{1v}^{1/2}\boldsymbol{G}_v & 0 & \boldsymbol{I} & * \\ \boldsymbol{Q}_{2v}^{1/2}\boldsymbol{W}_v & 0 & 0 & \boldsymbol{I} \end{pmatrix} \succ 0 \tag{3.22}$$

are feasible for all $v, w \in \mathbb{M}$. Furthermore, the switched terminal weighting matrices and switched feedback matrices are given by $\boldsymbol{Q}_{0v} = \boldsymbol{Z}_v^{-1}$ and $\boldsymbol{K}_v = \boldsymbol{W}_v\boldsymbol{G}_v^{-1}$ for all $v \in \mathbb{M}$.

PROOF. Assume that the LMIs (3.22) are feasible. Then $\boldsymbol{G}_v^T + \boldsymbol{G}_v - \boldsymbol{Z}_v \succ \boldsymbol{0}$ or equivalently $\boldsymbol{G}_v^T + \boldsymbol{G}_v \succ \boldsymbol{Z}_v \succ \boldsymbol{0}$ holds which follows from the Schur complement of (3.22). Hence, $\boldsymbol{G}_v$ is of full rank which is required to use $\boldsymbol{G}_v$ in a congruence transformation. Furthermore, as $\boldsymbol{Z}_v \succ \boldsymbol{0}$, also

$$(\boldsymbol{Z}_v - \boldsymbol{G}_v)^T \boldsymbol{Z}_v^{-1} (\boldsymbol{Z}_v - \boldsymbol{G}_v) \succ 0 \tag{3.23}$$

holds since an inversion and a congruence transformation do not affect definiteness. Equation (3.23) is equivalent to

$$\boldsymbol{G}_v^T \boldsymbol{Z}_v^{-1} \boldsymbol{G}_v \succ \boldsymbol{G}_v^T + \boldsymbol{G}_v - \boldsymbol{Z}_v. \tag{3.24}$$

Therefore, (3.22) implies

$$\begin{pmatrix} \boldsymbol{G}_v^T \boldsymbol{Z}_v^{-1} \boldsymbol{G}_v & * & * & * \\ \boldsymbol{A}_v\boldsymbol{G}_v - \boldsymbol{B}_v\boldsymbol{W}_v & \boldsymbol{Z}_w & * & * \\ \boldsymbol{Q}_{1v}^{1/2}\boldsymbol{G}_v & 0 & \boldsymbol{I} & * \\ \boldsymbol{Q}_{2v}^{1/2}\boldsymbol{W}_v & 0 & 0 & \boldsymbol{I} \end{pmatrix} \succ 0. \tag{3.25}$$

Substituting $\boldsymbol{W}_v = \boldsymbol{K}_v \boldsymbol{G}_v$ into (3.25) and pre-/post-multiplying by $\text{diag}(\boldsymbol{Z}_v \boldsymbol{G}_v^{-T}, \boldsymbol{I}, \boldsymbol{I}, \boldsymbol{I})$ and $\text{diag}(\boldsymbol{G}_v^{-1} \boldsymbol{Z}_v, \boldsymbol{I}, \boldsymbol{I}, \boldsymbol{I})$ leads to

$$\begin{pmatrix} \boldsymbol{Z}_v & * & * & * \\ \boldsymbol{A}_v \boldsymbol{Z}_v - \boldsymbol{B}_v \boldsymbol{K}_v \boldsymbol{Z}_v & \boldsymbol{Z}_w & * & * \\ \boldsymbol{Q}_{1v}^{1/2} \boldsymbol{Z}_v & 0 & \boldsymbol{I} & * \\ \boldsymbol{Q}_{2v}^{1/2} \boldsymbol{K}_v \boldsymbol{Z}_v & 0 & 0 & \boldsymbol{I} \end{pmatrix} \succ 0. \tag{3.26}$$

Applying the Schur complement to (3.26) results in

$$\boldsymbol{Z}_v - (\boldsymbol{A}_v \boldsymbol{Z}_v - \boldsymbol{B}_v \boldsymbol{K}_v \boldsymbol{Z}_v)^T \boldsymbol{Z}_w^{-1} (\boldsymbol{A}_v \boldsymbol{Z}_v - \boldsymbol{B}_v \boldsymbol{K}_v \boldsymbol{Z}_v) - \boldsymbol{Z}_v \boldsymbol{Q}_{1v} \boldsymbol{Z}_v - \boldsymbol{Z}_v \boldsymbol{K}_v^T \boldsymbol{Q}_{2v} \boldsymbol{K}_v \boldsymbol{Z}_v \succ 0. \tag{3.27}$$

Substituting $\boldsymbol{Z}_v = \boldsymbol{Q}_{0v}^{-1}$ into (3.27) and pre-/post-multiplying by $\boldsymbol{Q}_{0v}$ leads to (3.21), proving sufficiency. Necessity can be proved by extending the proof of Theorem 2 in [DRI02]. This proof is skipped for brevity. □

Remark 3.11. If only a common terminal weighting matrix $\boldsymbol{Q}_0$ should be considered for enforcing global uniform asymptotic stability of the PWL closed-loop system (3.8), then the LMIs (3.22) have to be solved with $\boldsymbol{Z}_v = \boldsymbol{Z}_w = \boldsymbol{Z}$. The common terminal weighting matrix then follows from $\boldsymbol{Q}_0 = \boldsymbol{Z}^{-1}$. Note that switched feedback matrices $\boldsymbol{K}_v$ can still be considered. This is enabled via the slack variable $\boldsymbol{G}_v$ which separates $\boldsymbol{Z}$ and $\boldsymbol{K}_v$. Without this slack variable, products of $\boldsymbol{Z}$ and $\boldsymbol{K}_v$ occur as can be observed in (3.26). An LMI formulation can then only be determined for a common feedback matrix $\boldsymbol{K} = \boldsymbol{K}_v$ using the substitution $\boldsymbol{W} = \boldsymbol{K}\boldsymbol{Z}$ in (3.26). The slack variable $\boldsymbol{G}_v$ in consequence allows to reduce the conservatism.

Remark 3.12. Similar LMIs as in Theorem 3.4 have been considered in different contexts, e.g. in [DRI02] for stability analysis and control synthesis for discrete-time switched linear systems, in [KBM96, CGM02, Mao03] for robust receding-horizon control or in [GKBM05] and [Laz06, Lemma 3.4.1] for stabilizing receding-horizon control of PWA systems. The particularities of the LMIs in Theorem 3.4 lie in using switched matrices $\boldsymbol{Z}_v$, $\boldsymbol{G}_v$ and $\boldsymbol{W}_v$ to minimize the conservatism and in avoiding an inversion of $\boldsymbol{Q}_{1v}$ to allow for a positive semidefinite $\boldsymbol{Q}_{1v}$. Note that a positive semidefinite $\boldsymbol{Q}_{1v}$ is, however, precluded in Theorem 3.3.

Recall that the switched terminal weighting matrix $\boldsymbol{Q}_{0j(k)}$ is only used for enforcing global uniform asymptotic stability. Since the control objective should be primarily influenced by the switched weighting matrices $\boldsymbol{Q}_{1j(k)}$ and $\boldsymbol{Q}_{2j(k)}$, it is desirable that the terminal cost term $\boldsymbol{x}^{*T}(k+N)\boldsymbol{Q}_{0j(k+N)}\boldsymbol{x}^*(k+N)$ affects the cost function only negligibly. Therefore, among all feasible solutions of (3.22) the one minimizing the terminal cost term should be selected.

First, minimizing the expected value of the terminal cost term assuming $\boldsymbol{x}^*(k+N)$ as Gaussian random variable with zero expected value and unit covariance matrix which corresponds to simultaneously minimizing the traces of all matrices $\boldsymbol{Q}_{0v}$ as shown in

Lemma A.1 can be considered. This leads to the LMI optimization problem

$$\begin{aligned} &\min_{\boldsymbol{Z}} -\log\det(\boldsymbol{Z}) \\ &\text{subject to} \begin{cases} (3.22) & \forall v, w \in \mathbb{M} \\ \boldsymbol{Z}_v - \boldsymbol{Z} \succeq \boldsymbol{0} & \forall v \in \mathbb{M} \end{cases} \end{aligned} \tag{3.28}$$

in the symmetric matrix $\boldsymbol{Z} \succ \boldsymbol{0}$. The proof is based on the following considerations:

Assume that $\boldsymbol{Z}_v - \boldsymbol{Z} \succeq \boldsymbol{0}$ is fulfilled for all $v \in \mathbb{M}$. Then

$$\begin{aligned} &\boldsymbol{Z}_v - \boldsymbol{Z} \succeq \boldsymbol{0} && \mid \boldsymbol{Z}^{-1}(\cdot)\boldsymbol{Z}_v^{-1} && (3.29a)\\ \Leftrightarrow\ &\boldsymbol{Z}^{-1} - \boldsymbol{Z}_v^{-1} \succeq \boldsymbol{0} && \mid \mathrm{tr}(\cdot) && (3.29b)\\ \Rightarrow\ &\mathrm{tr}(\boldsymbol{Z}^{-1} - \boldsymbol{Z}_v^{-1}) \geq 0 && && (3.29c)\\ \Leftrightarrow\ &\mathrm{tr}(\boldsymbol{Z}^{-1}) \geq \mathrm{tr}(\boldsymbol{Z}_v^{-1}) && \mid \boldsymbol{Z}_v^{-1} = \boldsymbol{Q}_{0v} && (3.29d)\\ \Leftrightarrow\ &\mathrm{tr}(\boldsymbol{Z}^{-1}) \geq \mathrm{tr}(\boldsymbol{Q}_{0v}). && && (3.29e) \end{aligned}$$

Minimizing $\mathrm{tr}(\boldsymbol{Z}^{-1})$ thus corresponds to simultaneously minimizing $\mathrm{tr}(\boldsymbol{Q}_{0v})$ for all $v \in \mathbb{M}$. Since $\mathrm{tr}(\boldsymbol{Z}^{-1})$ as an objective functions leads to a non-convex optimization problem, $\log\det(\boldsymbol{Z}^{-1}) = -\log\det(\boldsymbol{Z})$ is used as an objective function instead to obtain a convex optimization problem. This is motivated by the fact that $\mathrm{tr}(\cdot)$ corresponds to the sum of the eigenvalues and $\log\det(\cdot)$ corresponds to the sum of the logarithmized eigenvalues.

Second, minimizing the maximum value of the terminal cost term on the unit semi-hypersphere which corresponds to simultaneously minimizing the maximum eigenvalues of all matrices $\boldsymbol{Q}_{0v}$ as shown in Lemmas A.2 and 2.3 can be considered. This leads to the LMI optimization problem

$$\begin{aligned} &\max_{\lambda} \lambda \\ &\text{subject to} \begin{cases} (3.22) & \forall v, w \in \mathbb{M} \\ \boldsymbol{Z}_v - \lambda\boldsymbol{I} \succeq \boldsymbol{0} & \forall v \in \mathbb{M} \end{cases} \end{aligned} \tag{3.30}$$

in the scalar $\lambda > 0$. The proof is based on the following considerations:

Assume that $\boldsymbol{Z}_v - \lambda\boldsymbol{I} \succeq \boldsymbol{0}$ is fulfilled for all $v \in \mathbb{M}$. Then

$$\begin{aligned} &\boldsymbol{Z}_v - \lambda\boldsymbol{I} \succeq \boldsymbol{0} && \mid \boldsymbol{Z}_v^{-1}(\cdot) && (3.31a)\\ \Leftrightarrow\ &\boldsymbol{I} - \lambda\boldsymbol{Z}_v^{-1} \succeq \boldsymbol{0} && \mid \lambda^{-1}(\cdot) && (3.31b)\\ \Leftrightarrow\ &\lambda^{-1}\boldsymbol{I} - \boldsymbol{Z}_v^{-1} \succeq \boldsymbol{0} && \mid \boldsymbol{Z}_v^{-1} = \boldsymbol{Q}_{0v}, \lambda^{-1} = \lambda' && (3.31c)\\ \Leftrightarrow\ &\lambda'\boldsymbol{I} - \boldsymbol{Q}_{0v} \succeq \boldsymbol{0}. && && (3.31d) \end{aligned}$$

The scalar $\lambda' > 0$ is the maximum eigenvalue of the matrices $\boldsymbol{Q}_{0v}$ for all $v \in \mathbb{M}$, refer to [BEFB94, Section 2.2.2] for a comprehensive exposition on eigenvalue problems. Finally, maximizing λ corresponds to minimizing λ' due to $\lambda = \lambda'^{-1}$.

Closed-Loop Stability under the RRHCS Strategy

For the RRHCS strategy the relaxed value function $V_N(k)$ can not be directly used as a candidate Lyapunov function for stability analysis since it can not be defined uniquely. However, the boundaries $V_N^*(k) \leq V_N(k) \leq \alpha V_N^*(k)$ for all k and N can be determined, where the value function $V_N^*(k)$ is defined according to (3.14). Therefore, the worst case decrease of $V_N(k)$ can be considered:

$$V_N(k+1) - V_N(k) + \ell^*(k) \leq \alpha V_N^*(k+1) - V_N^*(k) + \ell^*(k) \overset{!}{<} 0. \tag{3.32}$$

Following then the ideas of the proof of Theorem 3.3, the step costs obviously do not cancel out when subtracting the value functions as in (3.18). This makes an LMI formulation particularly difficult.

In the following section several a posteriori stability criteria are proposed which can be used for both the RHCS and the RRHCS strategy.

3.3.2 A Posteriori Stability Analysis

A Posteriori Stability Analysis based on PWQ Lyapunov Functions

The first method for analyzing the stability of the switched system (1.3) under both the RHCS and the RRHCS strategy a posteriori is based on PWQ Lyapunov functions. The stability condition for the PWA system (1.4) given in Theorem 1.7 is reformulated for the PWL closed-loop system (3.8) to this end.

Proposition 3.5 *The PWL closed-loop system (3.8) is globally uniformly asymptotically stable if there exists a PWQ Lyapunov function*

$$V(\boldsymbol{x}(k)) = \boldsymbol{x}^T(k)\boldsymbol{S}_m\boldsymbol{x}(k) > 0 \quad \forall \boldsymbol{x}(k) \in \mathcal{X}_m \setminus \{\boldsymbol{0}\},\ m \in \hat{\mathbb{L}}_0 \tag{3.33}$$

whose difference along trajectories of the PWL closed-loop system is negative definite, i.e.

$$\begin{aligned}\Delta V(\boldsymbol{x}(k)) &= V(\boldsymbol{x}(k+1)) - V(\boldsymbol{x}(k)) \\ &= \boldsymbol{x}^T(k+1)\boldsymbol{S}_n\boldsymbol{x}(k+1) - \boldsymbol{x}^T(k)\boldsymbol{S}_m\boldsymbol{x}(k) \\ &= \boldsymbol{x}^T(k)\left(\widetilde{\boldsymbol{A}}_m^T\boldsymbol{S}_n\widetilde{\boldsymbol{A}}_m - \boldsymbol{S}_m\right)\boldsymbol{x}(k) \\ &< 0\end{aligned} \tag{3.34}$$

for all $\boldsymbol{x}(k) \in \mathcal{X}_m \setminus \{\boldsymbol{0}\}$ and all $\boldsymbol{x}(k+1) \in \mathcal{X}_n$ where $(m,n) \in \hat{\mathbb{L}}_0^2$ is a pair of consecutive controller indices, i.e. m relates to time instant k and n relates to time instant $k+1$.

PROOF. The proof follows similar lines as the proof of Theorem 1 in [FCMM02] and the discussion in [MFM00]. □

The conditions (3.33) and (3.34) must only hold in the regions $\mathcal{X}_m$ and $\mathcal{X}_n$ and not on the whole state space $\mathbb{R}^n$. Therefore, the *regionality* of the PWL closed-loop system is accounted for in the PWQ Lyapunov function. Finding such a PWQ Lyapunov function, however, is not a trivial task. Usually one must resort to more conservative conditions as outlined in Section 1.3.3.

A first approach to obtain a PWQ Lyapunov function consists in considering all pairs of consecutive controller indices while disregarding the regionality.

Theorem 3.6 (A Posteriori Stability Criterion) *If there exist symmetric matrices* $\boldsymbol{S}_m$ *with* $m \in \hat{\mathbb{L}}_0$ *such that the LMIs*

$$\boldsymbol{S}_m \succ 0 \tag{3.35a}$$

$$\widetilde{\boldsymbol{A}}_m^T \boldsymbol{S}_n \widetilde{\boldsymbol{A}}_m - \boldsymbol{S}_m \prec 0 \tag{3.35b}$$

are feasible for all pairs $(m, n) \in \hat{\mathbb{L}}_0^2$, *then the PWL closed-loop system (3.8) is globally uniformly asymptotically stable.*

PROOF. The proof follows similar lines as the discussion in [MFM00]. □

Remark 3.13. Note that (3.35) comprises $\hat{L}_0^2 + \hat{L}_0$ LMIs in $\hat{L}_0$ matrix variables.

Remark 3.14. Note that the PWQ Lyapunov function which underlies Theorem 3.6 is essentially a switched quadratic Lyapunov function (SQLF), cf. Theorem 1.6.

This criterion is clearly conservative because of disregarding the regionality. In the following a new method for including the regionality in Theorem 3.6 is presented. The cornerstone is

Lemma 3.7 *Consider the PWL closed-loop system (3.8). Furthermore, introduce the non-negative scalars* $\beta_{\hat{l}}, \gamma_{\hat{l}}$ *with* $\hat{l} \in \hat{\mathbb{L}}_0$ *and the matrix*

$$\underset{m,n}{\boldsymbol{\Delta}}(\beta_{\hat{l}}, \gamma_{\hat{l}}) = \sum_{\hat{l} \in \hat{\mathbb{L}}_0} \left[\beta_{\hat{l}}(\boldsymbol{P}_{\hat{l}} - \boldsymbol{P}_m) + \gamma_{\hat{l}} \widetilde{\boldsymbol{A}}_m^T (\boldsymbol{P}_{\hat{l}} - \boldsymbol{P}_n) \widetilde{\boldsymbol{A}}_m \right] \tag{3.36}$$

where $\boldsymbol{P}_{\hat{l}}$ *with* $\hat{l} \in \hat{\mathbb{L}}_0$ *are the DRE solutions defining the PWL closed-loop system (3.8).*

(i) If $\boldsymbol{x}(k) \in \mathcal{X}_m$ *and* $\boldsymbol{x}(k+1) = \widetilde{\boldsymbol{A}}_m \boldsymbol{x}(k) \in \mathcal{X}_n$,

then $\boldsymbol{x}^T(k) \underset{m,n}{\boldsymbol{\Delta}}(\beta_{\hat{l}}, \gamma_{\hat{l}}) \boldsymbol{x}(k) \geq 0$ *for arbitrary* $\beta_{\hat{l}}, \gamma_{\hat{l}} \geq 0$.

(ii) If $\boldsymbol{x}(k) \notin \mathcal{X}_m$ *or* $\boldsymbol{x}(k+1) = \widetilde{\boldsymbol{A}}_m \boldsymbol{x}(k) \notin \mathcal{X}_n$,

then there exist $\beta_{\hat{l}}, \gamma_{\hat{l}} \geq 0$ *such that* $\boldsymbol{x}^T(k) \underset{m,n}{\boldsymbol{\Delta}}(\beta_{\hat{l}}, \gamma_{\hat{l}}) \boldsymbol{x}(k) < 0$.

PROOF. Recall that according to (3.4) for all $\boldsymbol{x}(k) \in \mathcal{X}_m$ the inequalities

$$\boldsymbol{x}^T(k) \boldsymbol{P}_m \boldsymbol{x}(k) \leq \boldsymbol{x}^T(k) \boldsymbol{P}_{\hat{l}} \boldsymbol{x}(k) \tag{3.37}$$

are fulfilled for all $\hat{l} \in \hat{\mathbb{L}}_0$. Thereby the inequality

$$\boldsymbol{x}^T(k)\left(\sum_{\hat{l}\in\hat{\mathbb{L}}_0}\beta_{\hat{l}}(\boldsymbol{P}_{\hat{l}}-\boldsymbol{P}_m)\right)\boldsymbol{x}(k)\geq 0 \tag{3.38}$$

is implied for arbitrary scalars $\beta_{\hat{l}} \geq 0$. Conversely, for all $\boldsymbol{x}(k) \notin \mathcal{X}_m$ there exist $\hat{l} \in \hat{\mathbb{L}}_0$ for which the inequalities (3.37) do not hold and therefore there also exist $\beta_{\hat{l}} \geq 0$ such that the inequality (3.38) is not satisfied. Analogously, for all $\boldsymbol{x}(k+1) \in \mathcal{X}_n$ the inequality

$$\boldsymbol{x}^T(k+1)\left(\sum_{\hat{l}\in\hat{\mathbb{L}}_0}\gamma_{\hat{l}}(\boldsymbol{P}_{\hat{l}}-\boldsymbol{P}_n)\right)\boldsymbol{x}(k+1)\geq 0 \tag{3.39}$$

is fulfilled for arbitrary scalars $\gamma_{\hat{l}} \geq 0$. More strictly, for all $\boldsymbol{x}(k) \in \mathcal{X}_m$ such that $\boldsymbol{x}(k+1) = \widetilde{\boldsymbol{A}}_m\boldsymbol{x}(k) \in \mathcal{X}_n$ the inequality

$$\boldsymbol{x}^T(k)\left(\sum_{\hat{l}\in\hat{\mathbb{L}}_0}\gamma_{\hat{l}}\widetilde{\boldsymbol{A}}_m^T(\boldsymbol{P}_{\hat{l}}-\boldsymbol{P}_n)\widetilde{\boldsymbol{A}}_m\right)\boldsymbol{x}(k)\geq 0 \tag{3.40}$$

must hold for arbitrary $\gamma_{\hat{l}} \geq 0$. Conversely, for all $\boldsymbol{x}(k+1) \notin \mathcal{X}_n$ and $\widetilde{\boldsymbol{A}}_m\boldsymbol{x}(k) \notin \mathcal{X}_n$ respectively there exist $\gamma_{\hat{l}} \geq 0$ such that the inequality (3.39) and (3.40) respectively is not fulfilled.

By combining the terms contained in the inequalities (3.38) and (3.40), the matrix $\underset{m,n}{\boldsymbol{\Delta}}(\beta_{\hat{l}}, \gamma_{\hat{l}})$ according to (3.36) is obtained. □

A second approach to determine a PWQ Lyapunov function then consists in considering all pairs of consecutive controller indices while regarding the regionality using $\underset{m,n}{\boldsymbol{\Delta}}(\beta_{\hat{l}}, \gamma_{\hat{l}})$ and the S-procedure.

Theorem 3.8 (A Posteriori Stability Criterion Based on the S-Procedure) *If there exist symmetric matrices $\boldsymbol{S}_m$ with $m \in \hat{\mathbb{L}}_0$ and non-negative scalars $\alpha_{m\hat{l}}$, $\beta_{mn\hat{l}}$ and $\gamma_{mn\hat{l}}$ with $m, n, \hat{l} \in \hat{\mathbb{L}}_0$ such that the LMIs*

$$\boldsymbol{S}_m - \sum_{\hat{l}\in\hat{\mathbb{L}}_0}\alpha_{m\hat{l}}(\boldsymbol{P}_{\hat{l}}-\boldsymbol{P}_m) \succ \boldsymbol{0} \tag{3.41a}$$

$$\widetilde{\boldsymbol{A}}_m^T\boldsymbol{S}_n\widetilde{\boldsymbol{A}}_m - \boldsymbol{S}_m + \underset{m,n}{\boldsymbol{\Delta}}(\beta_{mn\hat{l}}, \gamma_{mn\hat{l}}) \prec \boldsymbol{0} \tag{3.41b}$$

are feasible for all pairs $(m, n) \in \hat{\mathbb{L}}_0^2$, then the PWL closed-loop system (3.8) is globally uniformly asymptotically stable.

PROOF. Using the S-procedure, the inequalities (3.38) and (3.40) can be included in the LMIs (3.35), leading to (3.41). If $\boldsymbol{x}(k) \in \mathcal{X}_m$ and $\boldsymbol{x}(k+1) \in \mathcal{X}_n$, then fulfillment of (3.41) implies fulfillment of (3.35), whereas if $\boldsymbol{x}(k) \notin \mathcal{X}_m$ or $\boldsymbol{x}(k+1) \notin \mathcal{X}_n$, then (3.41) is a relaxation of (3.35). □

Remark 3.15. Recalling that $\boldsymbol{P}_m$ are the DRE solutions defining the PWL closed-loop system (3.8) and only $\boldsymbol{S}_m$ are matrix variables, (3.41) comprises $\hat{L}_0^2 + \hat{L}_0$ LMIs in $\hat{L}_0$ matrix variables and $\hat{L}_0(\hat{L}_0 - 1) + 2\hat{L}_0^2(\hat{L}_0 - 1)$ scalar variables where specifically α_{mm}, β_{mnm} and γ_{mnn} may be chosen arbitrarily for all $m, n \in \hat{\mathbb{L}}_0$.

Remark 3.16. Both in Theorem 3.6 and 3.8 all pairs $(m, n) \in \hat{\mathbb{L}}_0^2$ are considered. Some pairs, however, may never occur or, in other words, may not be reachable. These pairs can be identified via a reachability analysis and omitted in Theorem 3.6 and 3.8 to reduce the complexity and conservatism. The reachability analysis, which is addressed in the forthcoming section, is inherently included in Theorem 3.8. Hence, for Theorem 3.8 only the complexity can be reduced. The conservatism of the a posteriori stability criteria can be classified as shown in Figure 3.2.

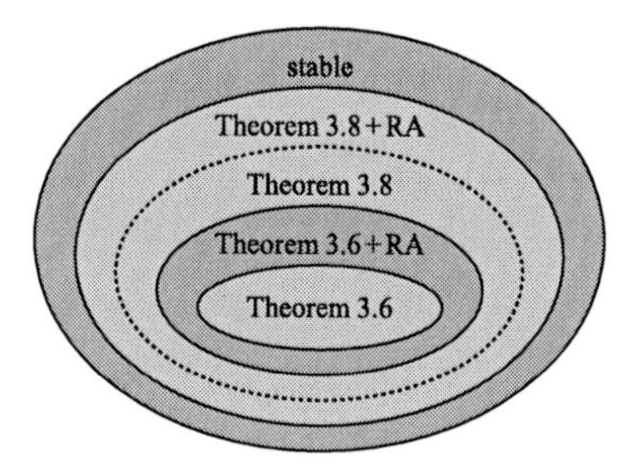

Figure 3.2: Conservatism of the a posteriori stability criteria given in Theorems 3.6 and 3.8 without/with preceding reachability analysis (RA)

Remark 3.17. For the existence of a PWQ Lyapunov function satisfying Theorem 3.6 or 3.8 the subsystems $\widetilde{\boldsymbol{A}}_m$ of the PWL closed-loop system (3.8) must be globally asymptotically stable for all $m \in \hat{\mathbb{L}}_0$ if no preceding reachability analysis is considered or for all $m \in \hat{\mathbb{L}}_0$ yielding a reachable pair $(m, m) \in \hat{\mathbb{L}}_0^2$ if a preceding reachability analysis is considered, cf. also Section 1.3.2. As a consequence, the subsystems $(\boldsymbol{A}_v, \boldsymbol{B}_v)$ of the switched system (1.3) must be stabilizable for all $v = \sigma(m)$.

A Posteriori Stability Analysis based on the Value Function

The second method for analyzing the stability of the switched system (1.3) under both the RHCS and the RRHCS strategy a posteriori is based on using the value function as a candidate Lyapunov function.

Theorem 3.9 (A Posteriori Stability Criterion based on the Value Function) *If the switched weighting matrices $\boldsymbol{Q}_{1j(k)}$ are strictly positive definite for all $j(k) \in \mathbb{M}$ and for each DRE solution $\boldsymbol{P}_m$, $m \in \hat{\mathbb{L}}_0$ there exist non-negative scalars $\xi_{\hat{l}}$, $\hat{l} \in \hat{\mathbb{L}}_0$ such that*

$$\sum_{\hat{l} \in \hat{\mathbb{L}}_0} \xi_{\hat{l}} = 1 \quad \text{and} \quad \sum_{\hat{l} \in \hat{\mathbb{L}}_0} \xi_{\hat{l}} \widetilde{\boldsymbol{A}}_m^T \boldsymbol{P}_{\hat{l}} \widetilde{\boldsymbol{A}}_m - \boldsymbol{P}_m \prec \boldsymbol{0}, \tag{3.42}$$

then the PWL closed-loop system (3.8) is globally uniformly asymptotically stable.

PROOF. The value function or, more precisely, the relaxed value function

$$V_N(\boldsymbol{x}(k)) = \min_{\hat{l} \in \hat{\mathbb{L}}_0} \boldsymbol{x}^T(k) \boldsymbol{P}_{\hat{l}} \boldsymbol{x}(k) \tag{3.43}$$

is positive definite, decrescent and radially unbounded since

$$\alpha_1 ||\boldsymbol{x}(k)||_2^2 \leq \boldsymbol{x}^T(k) \boldsymbol{P}_{\hat{l}} \boldsymbol{x}(k) \leq \alpha_2 ||\boldsymbol{x}(k)||_2^2 \quad \forall \hat{l} \in \hat{\mathbb{L}}_0 \quad \forall \boldsymbol{x}(k) \in \mathbb{R}^n \quad \forall k \in \mathbb{N}_0 \tag{3.44}$$

with

$$\alpha_1 = \min_{\hat{l} \in \hat{\mathbb{L}}_0} \lambda_{\min}(\boldsymbol{P}_{\hat{l}}) > 0, \quad \alpha_2 = \max_{\hat{l} \in \hat{\mathbb{L}}_0} \lambda_{\max}(\boldsymbol{P}_{\hat{l}}) > 0 \tag{3.45}$$

following from Lemma A.2 using that $\boldsymbol{P}_{\hat{l}} \succ \boldsymbol{0}$ for all $\hat{l} \in \hat{\mathbb{L}}_0$ due to requiring $\boldsymbol{Q}_{1j(k)} \succ \boldsymbol{0}$ for all $j(k) \in \mathbb{M}$, cf. also the DRE (2.3). The value function can therefore be used as a candidate Lyapunov function. The PWL closed-loop system (3.8) is globally uniformly asymptotically stable if the difference of the candidate Lyapunov function $V_N(\boldsymbol{x}(k))$ is negative definite along trajectories of the PWL closed-loop system, i.e. if

$$\begin{aligned} \Delta V_N(\boldsymbol{x}(k)) &= V_N(\boldsymbol{x}(k+1)) - V_N(\boldsymbol{x}(k)) \\ &= \min_{\hat{l} \in \hat{\mathbb{L}}_0} \boldsymbol{x}^T(k+1) \boldsymbol{P}_{\hat{l}} \boldsymbol{x}(k+1) - \min_{\hat{l} \in \hat{\mathbb{L}}_0} \boldsymbol{x}^T(k) \boldsymbol{P}_{\hat{l}} \boldsymbol{x}(k) \\ &< 0 \end{aligned} \tag{3.46}$$

for all $\boldsymbol{x}(k) \in \mathbb{R}^n \setminus \{\boldsymbol{0}\}$ and $k \in \mathbb{N}_0$.

Assume that the optimal controller index $m = \arg\min_{\hat{l} \in \hat{\mathbb{L}}_0} \boldsymbol{x}^T(k) \boldsymbol{P}_{\hat{l}} \boldsymbol{x}(k)$ is selected at time instant k, then

$$\Delta V_N(\boldsymbol{x}(k)) = \min_{\hat{l} \in \hat{\mathbb{L}}_0} \boldsymbol{x}^T(k+1) \boldsymbol{P}_{\hat{l}} \boldsymbol{x}(k+1) - \boldsymbol{x}^T(k) \boldsymbol{P}_m \boldsymbol{x}(k). \tag{3.47}$$

Using that for $\sum_{\hat{l} \in \hat{\mathbb{L}}_0} \xi_{\hat{l}} = 1$, $\xi_{\hat{l}} \geq 0$ with $\hat{l} \in \hat{\mathbb{L}}_0$ the inequality

$$\min_{\hat{l} \in \hat{\mathbb{L}}_0} \boldsymbol{x}^T(k+1) \boldsymbol{P}_{\hat{l}} \boldsymbol{x}(k+1) = \min_{\xi_{\hat{l}}} \sum_{\hat{l} \in \hat{\mathbb{L}}_0} \xi_{\hat{l}} \boldsymbol{x}^T(k+1) \boldsymbol{P}_{\hat{l}} \boldsymbol{x}(k+1) \leq \sum_{\hat{l} \in \hat{\mathbb{L}}_0} \xi_{\hat{l}} \boldsymbol{x}^T(k+1) \boldsymbol{P}_{\hat{l}} \boldsymbol{x}(k+1) \tag{3.48}$$

holds leads with $\boldsymbol{x}(k+1) = \widetilde{\boldsymbol{A}}_m \boldsymbol{x}(k)$ to

$$\begin{aligned} \Delta V_N(\boldsymbol{x}(k)) &\le \sum_{\hat{l} \in \hat{\mathbb{L}}_0} \xi_{\hat{l}} \boldsymbol{x}^T(k) \widetilde{\boldsymbol{A}}_m^T \boldsymbol{P}_{\hat{l}} \widetilde{\boldsymbol{A}}_m \boldsymbol{x}(k) - \boldsymbol{x}^T(k) \boldsymbol{P}_m \boldsymbol{x}(k) \\ &= \boldsymbol{x}^T(k) \left(\sum_{\hat{l} \in \hat{\mathbb{L}}_0} \xi_{\hat{l}} \widetilde{\boldsymbol{A}}_m^T \boldsymbol{P}_{\hat{l}} \widetilde{\boldsymbol{A}}_m - \boldsymbol{P}_m \right) \boldsymbol{x}(k). \end{aligned} \tag{3.49}$$

Therefore, if (3.42) is fulfilled, then also (3.46) is fulfilled, completing the proof. □

Remark 3.18. The a posteriori stability criterion given in Theorem 3.9 is very attractive due to its low computational complexity. Indeed, at most $\hat{L}_0$ LMI feasibility problems in $\hat{L}_0$ scalar variables must be solved. If a single $\boldsymbol{P}_m$ does not fulfill (3.42), then the evaluation can be stopped prematurely.

Remark 3.19. Contrary to Theorem 3.6 and 3.8, the global asymptotic stability of the subsystems $\widetilde{\boldsymbol{A}}_m$, $m \in \hat{\mathbb{L}}_0$ of the PWL closed-loop system (3.8) and thus the stabilizability of the subsystems $(\boldsymbol{A}_v, \boldsymbol{B}_v)$, $v = \sigma(m)$ of the switched system (1.3) is not a precondition for Theorem 3.9. This is achieved by using the switching law (3.5) between (3.46) and (3.47). Thus constrained instead of arbitrary switching is considered in Theorem 3.9.

Remark 3.20. Requiring $\boldsymbol{Q}_{1j(k)} \succ \boldsymbol{0}$ for all $j(k) \in \mathbb{M}$ to ensure $\boldsymbol{P}_{\hat{l}} \succ \boldsymbol{0}$ for all $\hat{l} \in \hat{\mathbb{L}}_0$ is a rather strict condition. Recalling the DRE (2.3), less strict conditions can be formulated. For example $\boldsymbol{K}_{l(N-1)}^T \boldsymbol{Q}_{2j(N-1)} \boldsymbol{K}_{l(N-1)} \succ \boldsymbol{0}$ for all $l(N-1) \in \mathbb{L}_{N-1}$ and $j(N-1) \in \mathbb{M}$ suffices to ensure $\boldsymbol{P}_{\hat{l}} \succ \boldsymbol{0}$ for all $\hat{l} \in \hat{\mathbb{L}}_0$. This condition can evidently not be enforced but easily checked.

Numerical Study

Example 3.2 Reconsider the setup of Example 2.3 in view of the RRHCS strategy (3.2). The corresponding PWL closed-loop systems (3.8) are analyzed a posteriori for stability using Theorems 3.6, 3.8 and 3.9. For Theorems 3.6 and 3.8 also the reachability is considered as suggested in Remark 3.16. The percentage of success as a function of the number of DRE solutions $|\hat{\mathbb{L}}_0|$ is plotted in Figure 3.3 for the system orders $\mathrm{n} \in \{2, 3\}$. For statistical significance, only results with an absolute frequency of the number of DRE solutions larger than five are regarded therein.

The percentage of success for Theorem 3.8 is considerably higher than for Theorem 3.6. Therefore, regarding the regionality is crucial. By considering a preceding reachability analysis, the percentage of success can be increased both for Theorem 3.6 and 3.8. The increase for Theorem 3.8, however, merely results from reducing the complexity whereby also numerical problems can be avoided. The percentage of success for Theorem 3.9 is close to that for Theorem 3.8 but declines with increasing $|\hat{\mathbb{L}}_0|$. Generally, the percentage of success degrades both for increasing $|\hat{\mathbb{L}}_0|$ and n. Overall, the classification of the conservatism according to Figure 3.2 is confirmed by the numerical study. The computation times are studied in Example 3.6 on page 57. □

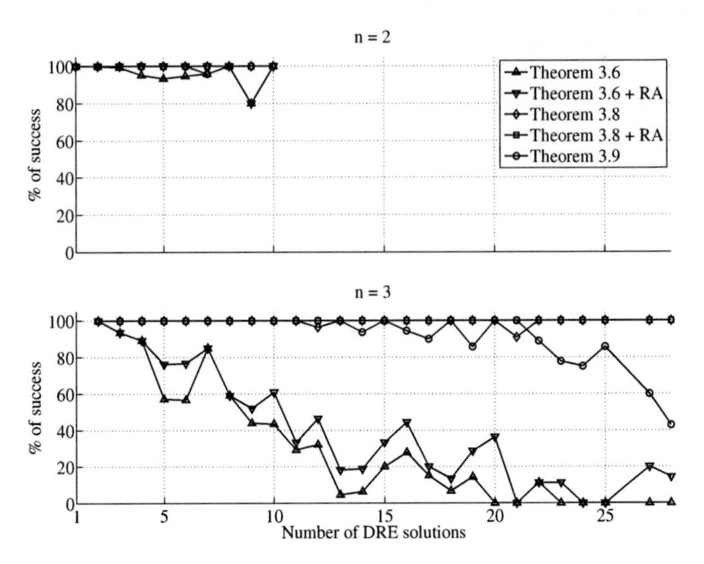

Figure 3.3: Percentage of success of the a posteriori stability analysis using Theorems 3.6 and 3.8 without/with preceding reachability analysis (RA) and Theorem 3.9

Modification of Problem 3.1 for Achieving Closed-Loop Stability

If the stability of the switched system (1.3) under the (R)RHCS strategy (3.3) can not be verified by the a posteriori stability criteria given in Theorems 3.6, 3.8 and 3.9, then Problem 3.1 can be modified in two ways:

First, the terminal weighting matrix $\boldsymbol{Q}_0$ may be "enlarged" which follows intuitively from Theorem 3.3 and the discussion at the end of Section 3.3.1. A reasonable initial guess for the terminal weighting matrix $\boldsymbol{Q}_0$ can be obtained from Theorems 3.3 and 3.4.

Second, the prediction horizon N can be increased. For time-invariant systems it has been shown [NP97, PN00, JH05, GMTT05, GR08, Grü09] that receding-horizon control is stabilizing if the prediction horizon is sufficiently large. Partially, estimates for the stabilizing prediction horizon have been formulated. The finite-horizon value function then becomes a Lyapunov function for the closed-loop system. Indeed, also Theorem 3.9 may be considered from this perspective. The results generally depend on controllability and observability conditions. Such controllability and observability conditions are difficult to obtain for switched systems. However, controllability and observability presumed, the fundamental principle also applies to switched systems.

Both approaches surely deserve a detailed analysis which is subject of future research.

The effectiveness of the approaches is illustrated by the following

Example 3.3 Hundred switched systems (1.3) with two third-order subsystems are generated (i.e. $M = 2$, n $= 3$). The subsystems are stabilizable and have random eigenvalues $\lambda_{1,2,3}$ fulfilling $\lambda_{1,2,3} \in \mathbb{C}$ with $|\lambda_{1,2,3}| \leq 4$ and $\lambda_{1,2,3} \notin \mathbb{R}^{-}$. For each switched system, Problem 3.1 is solved for $\boldsymbol{Q}_{1j} = \boldsymbol{I}$, $\boldsymbol{Q}_{2j} = 0 \; \forall j \in \mathbb{M}$, $\alpha = 1.01$ and either $N = 4$ and $r\boldsymbol{Q}_0$ with $r \in \{0, 1, 2\}$ and $\boldsymbol{Q}_0$ resulting from the LMI optimization problem (3.28) or $N \in \{2, \ldots, 6\}$ and $\boldsymbol{Q}_0 = \boldsymbol{0}$. The associated PWL closed-loop systems (3.8) are then analyzed for stability based on Theorems 3.8 and 3.9. The percentage of success is depicted in Figure 3.4.

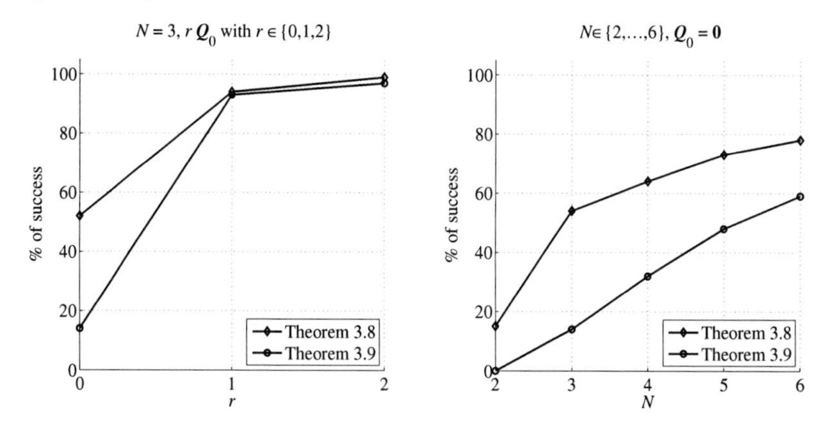

Figure 3.4: Percentage of success of the stability analysis in dependence of the terminal weighting matrix $\boldsymbol{Q}_0$ (left) and the prediction horizon N (right)

Obviously, the percentage of success increases with both the terminal weighting matrix $\boldsymbol{Q}_0$ and the prediction horizon N under both Theorem 3.8 and 3.9. □

In the following section a methodology for analyzing the reachability between the regions (3.4) is presented.

3.4 Reachability Analysis

The reachability analysis addresses the reachability between two regions $\mathcal{X}_m$ and $\mathcal{X}_n$ of the undisturbed PWL closed-loop system (3.8) for a pair of consecutive controller indices $(m, n) \in \hat{\mathbb{L}}_0^2$. From an analysis perspective it is more viable to consider the non-reachability:

Definition 3.10 (Non-Reachability) *Region $\mathcal{X}_n$ is not reachable from region $\mathcal{X}_m$ (i.e.*

$\mathcal{X}_m \nrightarrow \mathcal{X}_n$) iff $\boldsymbol{x}(k+1) = \widetilde{\boldsymbol{A}}_m \boldsymbol{x}(k) \notin \mathcal{X}_n$ for all $\boldsymbol{x}(k) \in \mathcal{X}_m$. Otherwise region $\mathcal{X}_n$ is reachable from region $\mathcal{X}_m$ (i.e. $\mathcal{X}_m \rightarrow \mathcal{X}_n$).

Based on Lemma 3.7 a sufficient non-reachability criterion can be formulated:

Theorem 3.11 (Non-Reachability Criterion) *If the inequality*

$$\underset{m,n}{\boldsymbol{\Delta}}(\beta_{\hat{l}}, \gamma_{\hat{l}}) \prec \boldsymbol{0} \tag{3.50}$$

holds for some non-negative scalars $\beta_{\hat{l}}, \gamma_{\hat{l}}$ with $\hat{l} \in \hat{\mathbb{L}}_0$, then $\mathcal{X}_m \nrightarrow \mathcal{X}_n$.

PROOF. If it is possible to find non-negative scalars $\beta_{\hat{l}}, \gamma_{\hat{l}}$ such that for all $\boldsymbol{x}(k) \in \mathbb{R}^{\mathrm{n}}$ the inequality $\boldsymbol{x}^T(k) \underset{m,n}{\boldsymbol{\Delta}}(\beta_{\hat{l}}, \gamma_{\hat{l}}) \boldsymbol{x}(k) < 0$ holds, then $\boldsymbol{x}(k) \notin \mathcal{X}_m$ or $\widetilde{\boldsymbol{A}}_m \boldsymbol{x}(k) \notin \mathcal{X}_n$ for any $\boldsymbol{x}(k) \in \mathbb{R}^{\mathrm{n}}$, therefore $\mathcal{X}_m \nrightarrow \mathcal{X}_n$. □

Remark 3.21. Inequality (3.50) is an LMI in $2(\hat{L}_0 - 1)$ scalar variables $\beta_{\hat{l}} > 0, \hat{l} \in \hat{\mathbb{L}}_0 \setminus \{m\}$ and $\gamma_{\hat{l}} > 0, \hat{l} \in \hat{\mathbb{L}}_0 \setminus \{n\}$ where β_m, γ_n may be chosen arbitrarily.

For a complete reachability analysis of the undisturbed PWL closed-loop system (3.8), the non-reachability criterion must be evaluated for all $(m,n) \in \mathbb{L}_0^2$, thus $\hat{L}_0^2$ LMI feasibility problems must be solved.

Remark 3.22. By applying the non-reachability criterion for all $(m,n) \in \hat{\mathbb{L}}_0^2$, a binary reachability matrix $\boldsymbol{R}$ with

$$R_{mn} = \begin{cases} 1 & \text{if } \mathcal{X}_m \rightarrow \mathcal{X}_n \\ 0 & \text{if } \mathcal{X}_m \nrightarrow \mathcal{X}_n \end{cases} \tag{3.51}$$

can be constructed. Furthermore, the reachability ratio $\rho \in (0, 1]$ can be defined as

$$\rho = \frac{1}{\hat{L}_0^2} \sum_{m \in \hat{\mathbb{L}}_0} \sum_{n \in \hat{\mathbb{L}}_0} R_{mn}. \tag{3.52}$$

Remark 3.23. Note that the non-reachability criterion applies to the PWL closed-loop system (3.8) without disturbances. For a PWL closed-loop system under disturbances the reachability analysis and the a posteriori analysis utilizing a preceding reachability analysis as proposed in Remark 3.16 must be reconsidered.

Example 3.4 Reconsider the setup of Example 3.1. The region reachability is analyzed utilizing the non-reachability criterion proposed in Theorem 3.11, leading to the reachability matrix

$$\boldsymbol{R} = \begin{pmatrix} 1 & 1 & 1 & 0 \\ 1 & 0 & 0 & 1 \\ 1 & 0 & 0 & 0 \\ 1 & 1 & 0 & 0 \end{pmatrix} \tag{3.53}$$

and the reachability ratio $\rho = 0.5$. □

Example 3.5 Reconsider the setup of Examples 2.3 and 3.2. The reachability between the regions of the PWL closed-loop systems (3.8) is analyzed via the non-reachability criterion given in Theorem 3.11. The mean of the reachability ratio ρ as a function of the number of DRE solutions $|\hat{\mathbb{L}}_0|$ is plotted in Figure 3.5 for the system orders $\mathrm{n} \in \{2, 3\}$.

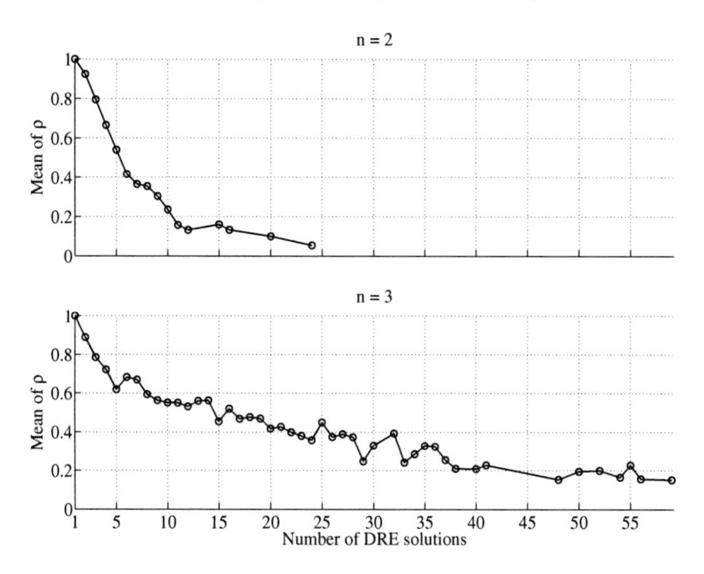

Figure 3.5: Mean of the reachability ratio ρ

Obviously, the mean of the reachability ratio decreases as $|\hat{\mathbb{L}}_0|$ increases. Moreover, the rate of decrease for $\mathrm{n} = 2$ is larger than for $\mathrm{n} = 3$. The computation times are discussed in the following Example 3.6. □

Example 3.6 The mean computation times for relaxed dynamic programming (see Example 2.3), the a posteriori stability analysis (see Example 3.2) and the reachability analysis (see Example 3.5) are plotted in Figure 3.6 on page 59 as a function of the number of DRE solutions $|\hat{\mathbb{L}}_0|$.

The mean computation times for evaluating Theorem 3.6 are considerably smaller than for evaluating Theorem 3.8, both with and without preceding reachability analysis. This is due the smaller number of variables involved in Theorem 3.6, see Remarks 3.13 and 3.15. By a preceding reachability analysis, the computation times for evaluating Theorems 3.6 and Theorem 3.8 can be reduced. The computation times for the reachability analysis must however be added. The total computation times can therefore be reduced only for large $|\hat{\mathbb{L}}_0|$ as can be observed from Figure 3.6 for $\mathrm{n} = 3$ regarding the logarithmic scale. On the other hand, as can be seen in Figure 3.6, Theorem 3.8 could not be evaluated for $|\hat{\mathbb{L}}_0| > 41$ without a preceding reachability analysis due to solver problems. From this point of view, a preceding reachability analysis may reduce the computational burden. The mean computation time for evaluating Theorem 3.9 is comparatively small, making Theorem 3.9 a good compromise between complexity and conservatism. The mean computation time for the reachability analysis is usually larger than for the a posteriori stability analysis. The obtained mean computation times clearly comply with Remarks 3.13, 3.15, 3.18 and 3.21. Overall, the mean computation times are acceptable even for large $|\hat{\mathbb{L}}_0|$. □

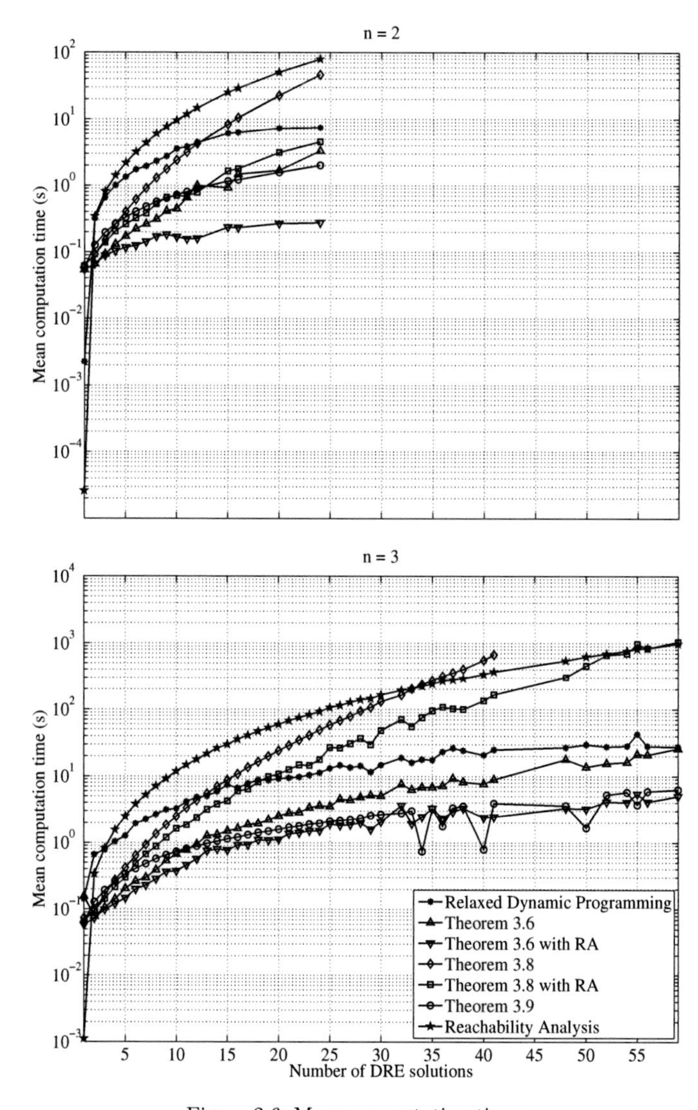

Figure 3.6: Mean computation times

4 Periodic Control and Scheduling

4.1 Offline Scheduling

4.1.1 Problem Formulation

Definition 4.1 *A switching sequence $j(k)$ with $k \in \mathbb{N}_0$ is called p-periodic if*

$$j(k) = j(k+p) \quad \forall k \in \mathbb{N}_0. \tag{4.1}$$

Consider the discrete-time quadratic cost function

$$J_p = \sum_{k=0}^{\infty} \left[\boldsymbol{x}^T(k) \boldsymbol{Q}_{1j(k)} \boldsymbol{x}(k) + \boldsymbol{u}^T(k) \boldsymbol{Q}_{2j(k)} \boldsymbol{u}(k) \right] \tag{4.2}$$

where $j(k)$ is a p-periodic switching sequence, $\boldsymbol{Q}_{1j(k)} \succeq \boldsymbol{0}$ and $\boldsymbol{Q}_{2j(k)} \succ \boldsymbol{0}$ with $j(k) \in \mathbb{M}$ are symmetric switched weighting matrices and the time horizon is infinite.

Problem 4.2 *For the switched system (1.3) find a control sequence $\boldsymbol{u}^*(k)$ and a p-periodic switching sequence $j^*(k)$ with $k \in \mathbb{N}_0$ such that the cost function (4.2) is minimized, i.e.*

$$\min_{\substack{\boldsymbol{u}(k),\ k \in \mathbb{N}_0 \\ j(k),\ k \in \mathbb{N}_0}} J_p \qquad \text{subject to (1.3).} \tag{4.3}$$

Problem 4.2 can be decomposed into a periodic control subproblem and a periodic scheduling subproblem. The periodic control subproblem can be solved based on periodic control theory while the periodic scheduling subproblem can be tackled by exhaustive search. In the following section periodic control theory is reviewed. A comprehensive presentation can be found in [BCD91, BC09].

4.1.2 Periodic Control

Problem Formulation

Consider the finite-horizon optimization problem

$$\min_{\boldsymbol{u}(k_0),\ldots,\boldsymbol{u}(k_0+Np-1)} \boldsymbol{x}^T(k_0+Np)\boldsymbol{Q}_0\boldsymbol{x}(k_0+Np) + \sum_{k=k_0}^{k_0+Np-1} \left[\boldsymbol{x}^T(k)\boldsymbol{Q}_{1j(k)}\boldsymbol{x}(k)+\boldsymbol{u}^T(k)\boldsymbol{Q}_{2j(k)}\boldsymbol{u}(k)\right] \tag{4.4a}$$

$$\text{subject to} \begin{cases} \boldsymbol{x}(k+1) &= \boldsymbol{A}_{j(k)}\boldsymbol{x}(k) + \boldsymbol{B}_{j(k)}\boldsymbol{u}(k) \\ \boldsymbol{x}(k_0) &= \boldsymbol{x}_{k_0} \end{cases} \tag{4.4b}$$

where $j(k)$ is a p-periodic switching sequence, $\boldsymbol{Q}_0 \succeq \boldsymbol{0}$ is a symmetric terminal weighting matrix, $\boldsymbol{Q}_{1j(k)} \succeq \boldsymbol{0}$ and $\boldsymbol{Q}_{2j(k)} \succ \boldsymbol{0}$ with $j(k) \in \mathbb{M}$ are symmetric switched weighting matrices, Np is the time horizon and $k_0 \in \mathbb{N}_0$ is a fixed initial time. The transition to an infinite time horizon is studied at the end of this section.

The solution of optimization problem (4.4) follows directly from linear-quadratic control theory for time-varying systems, see e.g. [KS72, Section 6.4] or [ÅW90, Section 11.2], as

$$\begin{aligned} \boldsymbol{u}^*(k) &= -\left(\boldsymbol{Q}_{2j(k)} + \boldsymbol{B}_{j(k)}^T\boldsymbol{P}(k+1)\boldsymbol{B}_{j(k)}\right)^{-1}\boldsymbol{B}_{j(k)}^T\boldsymbol{P}(k+1)\boldsymbol{A}_{j(k)}\boldsymbol{x}(k) \\ &= -\boldsymbol{K}^*(k)\boldsymbol{x}(k) \end{aligned} \tag{4.5}$$

where $\boldsymbol{P}(k)$ is the solution of the periodically time-varying difference Riccati equation (DRE)

$$\begin{aligned} \boldsymbol{P}(k) = &\left(\boldsymbol{A}_{j(k)} - \boldsymbol{B}_{j(k)}\boldsymbol{K}^*(k)\right)^T\boldsymbol{P}(k+1)\left(\boldsymbol{A}_{j(k)} - \boldsymbol{B}_{j(k)}\boldsymbol{K}^*(k)\right) \\ &+ \boldsymbol{Q}_{1j(k)} + \boldsymbol{K}^{*T}(k)\boldsymbol{Q}_{2j(k)}\boldsymbol{K}^*(k) \end{aligned} \tag{4.6}$$

for $k = k_0 + Np - 1, \ldots, k_0$ with the terminal condition $\boldsymbol{P}(k_0 + Np) = \boldsymbol{Q}_0$. This DRE is commonly denoted as discrete-time periodic Riccati equation (DPRE). The minimum cost is given by

$$\boldsymbol{x}^T(k_0)\boldsymbol{P}(k_0)\boldsymbol{x}(k_0). \tag{4.7}$$

The solution $\boldsymbol{P}(k)$ of the DPRE (4.6) is generally not periodic unless an appropriate terminal condition called periodic generator is imposed. The periodic generator emulates the cost for an infinite time horizon, i.e. from $k = k_0+Np$ to $k = \infty$. Periodicity of $\boldsymbol{P}(k)$ is enforced in this way. The periodic generator can be determined by reformulating the periodically time-varying optimization problem (4.4) as a time-invariant optimization problem using the lifting procedure. To simplify lifting, a factorization of the cost function in (4.4a) is in order. Lifting of the cost function can be avoided thereby.

Factorization

Consider the factorizations

$$\boldsymbol{Q}_{1j(k)} = \boldsymbol{C}_{j(k)}^T\boldsymbol{C}_{j(k)} \tag{4.8a}$$

$$\boldsymbol{Q}_{2j(k)} = \boldsymbol{L}_{j(k)}^T\boldsymbol{L}_{j(k)} \tag{4.8b}$$

with $\boldsymbol{C}_{j(k)} \in \mathbb{R}^{\mathrm{n}\times\mathrm{n}}$ symmetric and positive semidefinite and $\boldsymbol{L}_{j(k)} \in \mathbb{R}^{\mathrm{m}\times\mathrm{m}}$ symmetric and positive definite. Such factorizations can be determined using the Cholesky decomposition, see [GV96, Section 4.2]. Since $\boldsymbol{L}_{j(k)}$ is positive definite, it is also invertible. Based on the factorizations (4.8), a new control vector

$$\check{\boldsymbol{u}}(k) = \boldsymbol{L}_{j(k)}\boldsymbol{u}(k) \tag{4.9}$$

and input matrix

$$\check{\boldsymbol{B}}_{j(k)} = \boldsymbol{B}_{j(k)}\boldsymbol{L}_{j(k)}^{-1} \tag{4.10}$$

can be defined, leading to the factorized optimization problem

$$\min_{\check{\boldsymbol{u}}(k_0),\ldots,\check{\boldsymbol{u}}(k_0+Np-1)} \boldsymbol{x}^T(k_0+Np)\boldsymbol{Q}_0\boldsymbol{x}(k_0+Np) + \sum_{k=k_0}^{k_0+Np-1} \left[\boldsymbol{y}^T(k)\boldsymbol{y}(k) + \check{\boldsymbol{u}}^T(k)\check{\boldsymbol{u}}(k)\right] \tag{4.11a}$$

$$\text{subject to} \left\{\begin{array}{rcl} \boldsymbol{x}(k+1) &=& \boldsymbol{A}_{j(k)}\boldsymbol{x}(k) + \check{\boldsymbol{B}}_{j(k)}\check{\boldsymbol{u}}(k) \\ \boldsymbol{y}(k) &=& \boldsymbol{C}_{j(k)}\boldsymbol{x}(k) \\ \boldsymbol{x}(k_0) &=& \boldsymbol{x}_{k_0}. \end{array}\right. \tag{4.11b}$$

The factorized optimization problem (4.11) can now be lifted.

Lifting

The idea of lifting consists in downsampling the state vector $\boldsymbol{x}(k)$ of the periodically time-varying system (4.11b) with the period p, leading to the lifted state vector

$$\overline{\boldsymbol{x}}_{k_0}(\kappa) = \boldsymbol{x}(k_0+\kappa p) \tag{4.12}$$

with $\overline{\boldsymbol{x}}_{k_0}(\kappa) \in \mathbb{R}^{\mathrm{n}}$ and $\kappa = \lfloor (k-k_0)/p \rfloor$ denoting the time instant of the lifted system. The lifted control and output vector

$$\overline{\boldsymbol{u}}_{k_0}(\kappa) = \begin{pmatrix}\check{\boldsymbol{u}}^T(k_0+\kappa p) & \check{\boldsymbol{u}}^T(k_0+\kappa p+1) & \cdots & \check{\boldsymbol{u}}^T(k_0+(\kappa+1)p-1)\end{pmatrix}^T \tag{4.13a}$$

$$\overline{\boldsymbol{y}}_{k_0}(\kappa) = \begin{pmatrix}\boldsymbol{y}^T(k_0+\kappa p) & \boldsymbol{y}^T(k_0+\kappa p+1) & \cdots & \boldsymbol{y}^T(k_0+(\kappa+1)p-1)\end{pmatrix}^T \tag{4.13b}$$

with $\overline{\boldsymbol{u}}_{k_0}(\kappa) \in \mathbb{R}^{\mathrm{pm}}$ and $\overline{\boldsymbol{y}}_{k_0}(\kappa) \in \mathbb{R}^{\mathrm{pn}}$ follow by collecting the control vectors $\check{\boldsymbol{u}}(k)$ and output vectors $\boldsymbol{y}(k)$ of the periodically time-varying system (4.11b) over the period p. The lifted system, input, output and feedthrough matrix are obtained by solving the periodically time-varying system (4.11b) over the period p as

$$\overline{\boldsymbol{A}}_{k_0} = \boldsymbol{\Psi}(k_0) \tag{4.14a}$$

$$\overline{\boldsymbol{B}}_{k_0} = \begin{pmatrix}\boldsymbol{\Phi}(k_0+p,k_0+1)\check{\boldsymbol{B}}_{j(k_0)} & \boldsymbol{\Phi}(k_0+p,k_0+2)\check{\boldsymbol{B}}_{j(k_0+1)} & \cdots & \check{\boldsymbol{B}}_{j(k_0+p-1)}\end{pmatrix} \tag{4.14b}$$

$$\overline{\boldsymbol{C}}_{k_0} = \begin{pmatrix}\boldsymbol{C}_{j(k_0)}^T & \boldsymbol{\Phi}^T(k_0+1,k_0)\boldsymbol{C}_{j(k_0+1)}^T & \cdots & \boldsymbol{\Phi}^T(k_0+p-1,k_0)\boldsymbol{C}_{j(k_0+p-1)}^T\end{pmatrix}^T \tag{4.14c}$$

$$\overline{\boldsymbol{D}}_{k_0} = \left(\left(\overline{\boldsymbol{D}}_{k_0}\right)_{rs}\right), \quad r,s = 1,2,\ldots,p \tag{4.14d}$$

$$\left(\overline{\boldsymbol{D}}_{k_0}\right)_{rs} = \left\{\begin{array}{ll} \boldsymbol{0}, & r \leq s \\ \boldsymbol{C}_{j(k_0+r-1)}\boldsymbol{\Phi}(k_0+r-1,k_0+s)\check{\boldsymbol{B}}_{j(k_0+s-1)}, & r > s \end{array}\right.$$

with $\overline{\boldsymbol{A}}_{k_0} \in \mathbb{R}^{\mathrm{n} \times \mathrm{n}}$, $\overline{\boldsymbol{B}}_{k_0} \in \mathbb{R}^{\mathrm{n} \times \mathrm{pm}}$, $\overline{\boldsymbol{C}}_{k_0} \in \mathbb{R}^{\mathrm{pn} \times \mathrm{n}}$, $\overline{\boldsymbol{D}}_{k_0} \in \mathbb{R}^{\mathrm{pn} \times \mathrm{pm}}$ and $\boldsymbol{\Phi}(k, k_0)$ denoting the transition matrix of the periodically time-varying system (4.11b) given by

$$\boldsymbol{\Phi}(k, k_0) = \begin{cases} \boldsymbol{I}, & k = k_0 \\ \boldsymbol{A}_{j(k-1)} \boldsymbol{A}_{j(k-2)} \cdots \boldsymbol{A}_{j(k_0)}, & k > k_0. \end{cases} \tag{4.15}$$

The transition matrix over the period p

$$\boldsymbol{\Psi}(k_0) = \boldsymbol{\Phi}(k_0 + p, k_0) \tag{4.16}$$

is denoted as monodromy matrix. The eigenvalues of the monodromy matrix are called characteristic multipliers and are independent of the initial time k_0 [BC09, Section 3.1.2]. Consequently, the periodically time-varying system (4.11b) is globally asymptotically stable iff the characteristic multipliers of the associated monodromy matrix $\boldsymbol{\Psi}(k_0)$ are within the open unit disc of the complex plane [BC09, Section 3.3.1].

The lifting procedure is illustrated in Figure 4.1 for $p = 3$ and $k_0 = 0$ or $k_0 = 1$. Note that for conciseness only lifting of the state vector and the system matrix is visualized.

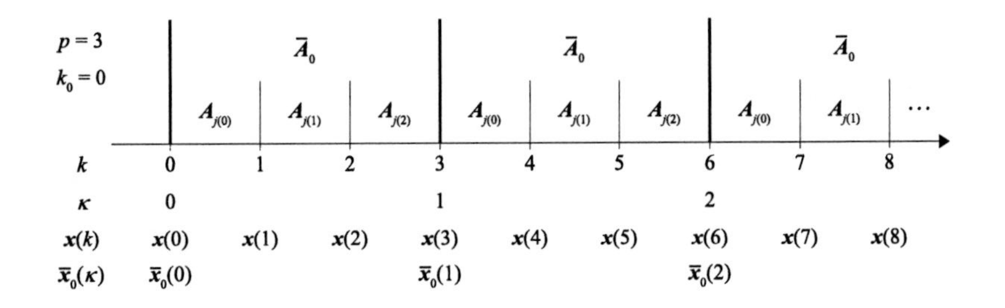

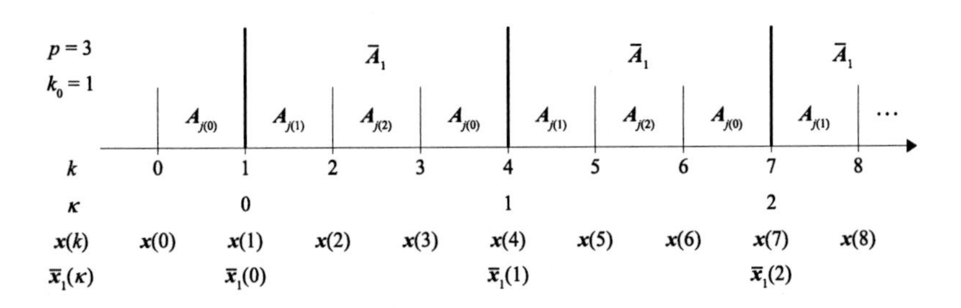

Figure 4.1: Illustration of the lifting procedure

The figure reveals that the initial time k_0 may also be interpreted as the starting index of the p-periodic switching sequence $j(k)$.

The lifted optimization problem can now be formulated as

$$\min_{\overline{\boldsymbol{u}}_{k_0}(0),\ldots,\overline{\boldsymbol{u}}_{k_0}(N-1)} \overline{\boldsymbol{x}}_{k_0}^T(N)\boldsymbol{Q}_0\overline{\boldsymbol{x}}_{k_0}(N) + \sum_{\kappa=0}^{N-1}\left[\overline{\boldsymbol{y}}_{k_0}^T(\kappa)\overline{\boldsymbol{y}}_{k_0}(\kappa) + \overline{\boldsymbol{u}}_{k_0}^T(\kappa)\overline{\boldsymbol{u}}_{k_0}(\kappa)\right] \tag{4.17a}$$

$$\text{subject to}\left\{\begin{array}{rcl}\overline{\boldsymbol{x}}_{k_0}(\kappa+1) &=& \overline{\boldsymbol{A}}_{k_0}\overline{\boldsymbol{x}}_{k_0}(\kappa) + \overline{\boldsymbol{B}}_{k_0}\overline{\boldsymbol{u}}_{k_0}(\kappa)\\ \overline{\boldsymbol{y}}_{k_0}(\kappa) &=& \overline{\boldsymbol{C}}_{k_0}\overline{\boldsymbol{x}}_{k_0}(\kappa) + \overline{\boldsymbol{D}}_{k_0}\overline{\boldsymbol{u}}_{k_0}(\kappa)\\ \overline{\boldsymbol{x}}_{k_0}(0) &=& \boldsymbol{x}_{k_0}\end{array}\right. \tag{4.17b}$$

or after resubstituting the output equation $\overline{\boldsymbol{y}}_{k_0}(\kappa) = \overline{\boldsymbol{C}}_{k_0}\overline{\boldsymbol{x}}_{k_0}(\kappa) + \overline{\boldsymbol{D}}_{k_0}\overline{\boldsymbol{u}}_{k_0}(\kappa)$ into the cost function in (4.17a) as

$$\min_{\overline{\boldsymbol{u}}_{k_0}(0),\ldots,\overline{\boldsymbol{u}}_{k_0}(N-1)} \overline{\boldsymbol{x}}_{k_0}^T(N)\boldsymbol{Q}_0\overline{\boldsymbol{x}}_{k_0}(N) + \tag{4.18a}$$

$$\sum_{\kappa=0}^{N-1}\left[\overline{\boldsymbol{x}}_{k_0}^T(\kappa)\overline{\boldsymbol{Q}}_{1k_0}\overline{\boldsymbol{x}}_{k_0}(\kappa) + 2\overline{\boldsymbol{x}}_{k_0}^T(\kappa)\overline{\boldsymbol{Q}}_{12k_0}\overline{\boldsymbol{u}}_{k_0}(\kappa) + \overline{\boldsymbol{u}}_{k_0}^T(\kappa)\overline{\boldsymbol{Q}}_{2k_0}\overline{\boldsymbol{u}}_{k_0}(\kappa)\right]$$

$$\text{subject to}\left\{\begin{array}{rcl}\overline{\boldsymbol{x}}_{k_0}(\kappa+1) &=& \overline{\boldsymbol{A}}_{k_0}\overline{\boldsymbol{x}}_{k_0}(\kappa) + \overline{\boldsymbol{B}}_{k_0}\overline{\boldsymbol{u}}_{k_0}(\kappa)\\ \overline{\boldsymbol{x}}_{k_0}(0) &=& \boldsymbol{x}_{k_0}\end{array}\right. \tag{4.18b}$$

with $\overline{\boldsymbol{C}}_{k_0}^T\overline{\boldsymbol{C}}_{k_0} = \overline{\boldsymbol{Q}}_{1k_0} \in \mathbb{R}^{\mathrm{n}\times\mathrm{n}}$, $\overline{\boldsymbol{C}}_{k_0}^T\overline{\boldsymbol{D}}_{k_0} = \overline{\boldsymbol{Q}}_{12k_0} \in \mathbb{R}^{\mathrm{n}\times\mathrm{pm}}$ and $\overline{\boldsymbol{D}}_{k_0}^T\overline{\boldsymbol{D}}_{k_0} + \boldsymbol{I} = \overline{\boldsymbol{Q}}_{2k_0} \in \mathbb{R}^{\mathrm{pm}\times\mathrm{pm}}$. The optimization problem (4.18) is time-invariant. Hence, the solution directly follows from linear-quadratic control theory as

$$\begin{aligned}\overline{\boldsymbol{u}}_{k_0}^*(\kappa) &= -\left(\overline{\boldsymbol{Q}}_{2k_0} + \overline{\boldsymbol{B}}_{k_0}^T\overline{\boldsymbol{P}}_{k_0}(\kappa+1)\overline{\boldsymbol{B}}_{k_0}\right)^{-1}\left(\overline{\boldsymbol{B}}_{k_0}^T\overline{\boldsymbol{P}}_{k_0}(\kappa+1)\overline{\boldsymbol{A}}_{k_0} + \overline{\boldsymbol{Q}}_{12k_0}^T\right)\overline{\boldsymbol{x}}_{k_0}(\kappa)\\ &= -\overline{\boldsymbol{K}}_{k_0}^*(\kappa)\overline{\boldsymbol{x}}_{k_0}(\kappa)\end{aligned} \tag{4.19}$$

where $\overline{\boldsymbol{P}}_{k_0}(\kappa)$ is the solution of the time-invariant DRE

$$\begin{aligned}\overline{\boldsymbol{P}}_{k_0}(\kappa) =& \left(\overline{\boldsymbol{A}}_{k_0} - \overline{\boldsymbol{B}}_{k_0}\overline{\boldsymbol{K}}_{k_0}^*(\kappa)\right)^T\overline{\boldsymbol{P}}_{k_0}(\kappa+1)\left(\overline{\boldsymbol{A}}_{k_0} - \overline{\boldsymbol{B}}_{k_0}\overline{\boldsymbol{K}}_{k_0}^*(\kappa)\right)\\ &+ \overline{\boldsymbol{Q}}_{1k_0} - \overline{\boldsymbol{K}}_{k_0}^{*\,T}(\kappa)\overline{\boldsymbol{Q}}_{12k_0}^T - \overline{\boldsymbol{Q}}_{12k_0}\overline{\boldsymbol{K}}_{k_0}^*(\kappa) + \overline{\boldsymbol{K}}_{k_0}^{*\,T}(\kappa)\overline{\boldsymbol{Q}}_{2k_0}\overline{\boldsymbol{K}}_{k_0}^*(\kappa)\end{aligned} \tag{4.20}$$

for $\kappa = N-1,\ldots,0$ with the terminal condition $\overline{\boldsymbol{P}}_{k_0}(N) = \boldsymbol{Q}_0$. The minimum cost is given by

$$\overline{\boldsymbol{x}}_{k_0}^T(0)\overline{\boldsymbol{P}}_{k_0}(0)\overline{\boldsymbol{x}}_{k_0}(0). \tag{4.21}$$

Relations between the Optimization Problems (4.4) and (4.18)

The relation between the symmetric positive semidefinite solutions of the DPRE (4.6) and the DRE (4.20) is given by

Lemma 4.3 ([BCD91, Lemma 6.4]) *If* $\overline{\boldsymbol{P}}_{k_0}(N) = \boldsymbol{P}(k_0+Np) = \boldsymbol{Q}_0$, *then*

$$\overline{\boldsymbol{P}}_{k_0}(\kappa) = \boldsymbol{P}(k_0+\kappa p) \quad \forall 0 \le \kappa \le N. \tag{4.22}$$

This relation follows from the coincidence of the optimization problems (4.4) and (4.18). Furthermore, the limiting behavior of the DRE (4.20) can be inferred from linear-quadratic (LQ) control theory, see [KS72, Theorem 6.31] and [Ber05a, Proposition 4.4.1]:

Proposition 4.4 *If $(\overline{\boldsymbol{A}}_{k_0}, \overline{\boldsymbol{B}}_{k_0})$ is a stabilizable pair, then for each positive semidefinite terminal condition $\overline{\boldsymbol{P}}_{k_0}(N)$ there exists a constant symmetric positive semidefinite matrix $\overline{\boldsymbol{P}}_{k_0}$ with*

$$\overline{\boldsymbol{P}}_{k_0} = \lim_{\kappa \to -\infty} \overline{\boldsymbol{P}}_{k_0}(\kappa). \tag{4.23}$$

The optimal control law is then given by

$$\begin{aligned} \overline{\boldsymbol{u}}^*_{k_0}(\kappa) &= -\left(\overline{\boldsymbol{Q}}_{2k_0} + \overline{\boldsymbol{B}}^T_{k_0}\overline{\boldsymbol{P}}_{k_0}\overline{\boldsymbol{B}}_{k_0}\right)^{-1}\left(\overline{\boldsymbol{B}}^T_{k_0}\overline{\boldsymbol{P}}_{k_0}\overline{\boldsymbol{A}}_{k_0} + \overline{\boldsymbol{Q}}^T_{12k_0}\right)\overline{\boldsymbol{x}}_{k_0}(\kappa) \\ &= -\overline{\boldsymbol{K}}^*_{k_0}\overline{\boldsymbol{x}}_{k_0}(\kappa) \end{aligned} \tag{4.24}$$

where $\overline{\boldsymbol{P}}_{k_0}$ is the unique solution of the algebraic Riccati equation (ARE)

$$\begin{aligned} \overline{\boldsymbol{P}}_{k_0} &= \left(\overline{\boldsymbol{A}}_{k_0} - \overline{\boldsymbol{B}}_{k_0}\overline{\boldsymbol{K}}^*_{k_0}\right)^T \overline{\boldsymbol{P}}_{k_0}\left(\overline{\boldsymbol{A}}_{k_0} - \overline{\boldsymbol{B}}_{k_0}\overline{\boldsymbol{K}}^*_{k_0}\right) \\ &\quad + \overline{\boldsymbol{Q}}_{1k_0} - \overline{\boldsymbol{K}}^{*\,T}_{k_0}\overline{\boldsymbol{Q}}^T_{12k_0} - \overline{\boldsymbol{Q}}_{12k_0}\overline{\boldsymbol{K}}^*_{k_0} + \overline{\boldsymbol{K}}^{*\,T}_{k_0}\overline{\boldsymbol{Q}}_{2k_0}\overline{\boldsymbol{K}}^*_{k_0}. \end{aligned} \tag{4.25}$$

The minimum cost is given by

$$\overline{\boldsymbol{x}}^T_{k_0}(0)\overline{\boldsymbol{P}}_{k_0}\overline{\boldsymbol{x}}_{k_0}(0). \tag{4.26}$$

Lemma 4.3 and Proposition 4.4 together implicate that $\overline{\boldsymbol{P}}_{k_0}$ is the periodic generator of the DPRE (4.6), yielding

Proposition 4.5 ([BCD91, Proposition 6.1]) *If $\boldsymbol{P}(k)$ is a symmetric periodic positive semidefinite (SPPS) solution of the DPRE (4.6), then $\overline{\boldsymbol{P}}_{k_0} = \boldsymbol{P}(k_0)$ is a positive semidefinite solution of the ARE (4.25). Conversely, if $\overline{\boldsymbol{P}}_{k_0}$ is a positive semidefinite solution of (4.25), then the solution of (4.6) with $\boldsymbol{P}(k_0) = \overline{\boldsymbol{P}}_{k_0}$ is SPPS.*

The SPPS solution of the DPRE (4.6) is thus given by

$$\boldsymbol{P}(k) = \overline{\boldsymbol{P}}_{k \bmod p} \tag{4.27}$$

and can be obtained by solving the ARE (4.25) for $k_0 = 0, \ldots, p-1$ using standard methods. The optimal p-periodic feedback matrices, in the following denoted as periodic linear-quadratic regulator (PLQR), result as

$$\boldsymbol{K}^*(k) = \boldsymbol{L}^{-1}_{j(k)}\left(\overline{\boldsymbol{K}}^*_{k \bmod p}\right)_{rs} \quad \text{for } 1 \leq r \leq \mathrm{m}, 1 \leq s \leq \mathrm{n}. \tag{4.28}$$

The minimum cost is given by

$$\boldsymbol{x}^T(k_0)\overline{\boldsymbol{P}}_{k_0}\boldsymbol{x}(k_0) = \boldsymbol{x}^T(k_0)\boldsymbol{P}(k_0)\boldsymbol{x}(k_0). \tag{4.29}$$

The design of a PLQR is summarized in Figure 4.2.

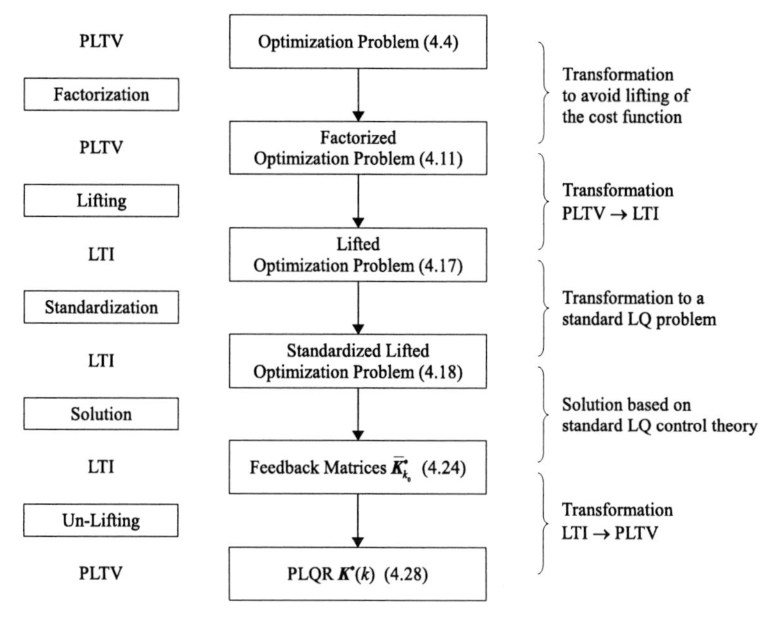

Figure 4.2: Design of a PLQR

Stability

The stability of the closed-loop system can be inferred from linear-quadratic control theory, see [KS72, Theorem 6.31], [Ber05a, Proposition 4.4.1] and furthermore [BCD91, Section 6.4.3] and [BCD88, Theorem 6]:

Proposition 4.6 *If $(\overline{A}_{k_0}, \overline{B}_{k_0})$ is a stabilizable pair and $(\overline{C}_{k_0}, \overline{A}_{k_0})$ is a detectable pair, then the closed-loop system*

$$x(k+1) = \left(A_{j(k)} - K^*(k) B_{j(k)}\right) x(k) \tag{4.30}$$

is globally asymptotically stable.

Remark 4.1. The stabilizability of $(\overline{A}_{k_0}, \overline{B}_{k_0})$ and the detectability of $(\overline{C}_{k_0}, \overline{A}_{k_0})$ are independent of the initial time k_0, see [BC09, Remark 6.2] and [BC00, Remark 8] and furthermore [BC09, Sections 4.4.1 and 4.4.2], [BB85b, Lemma 3], [BB85a, p. 919] and [Bit86, p. 160]. Thus, if stabilizability and detectability can be verified for a specific k_0, then stabilizability and detectability are verified for all $k_0 \in \mathbb{N}_0$.

Note that the results presented in this section can alternatively be derived based on dynamic programming, see [Ber01, Section 3.6, Exercise 3.12].

In summary, the PLQR (4.28) characterizes the p-periodic solution of the periodically time-varying optimization problem (4.4) for $N \to \infty$ and therefore also the solution of the periodic control subproblem of Problem 4.2.

In the following section the solution of Problem 4.2 comprising both the periodic control subproblem and the periodic scheduling subproblem is determined by combining PLQR and exhaustive search.

4.1.3 Solution based on Periodic Control and Exhaustive Search

For formulating the solution of Problem 4.2, some further definitions are in order.

Definition 4.7 *The set of p-periodic switching sequences is defined by*

$$\mathbb{J}_p = \{j(k) | j(k) = j(k+p) \ \forall k \in \mathbb{N}_0\}. \tag{4.31}$$

Definition 4.8 *A variable related to a particular p-periodic switching sequence $j(k) \in \mathbb{J}_p$ is marked by the subindex j, e.g.*

$$\boldsymbol{P}_j(k),\ \boldsymbol{K}_j(k),\ \overline{\boldsymbol{A}}_{k_0 j},\ \overline{\boldsymbol{B}}_{k_0 j},\ \overline{\boldsymbol{C}}_{k_0 j},\ \overline{\boldsymbol{D}}_{k_0 j},\ \overline{\boldsymbol{Q}}_{1k_0 j},\ \overline{\boldsymbol{Q}}_{12k_0 j},\ \overline{\boldsymbol{Q}}_{2k_0 j},\ \mathcal{X}_j,\ \xi_j. \tag{4.32}$$

Remark 4.2. The subindex j constitutes an indexation of the solution space $\mathbb{J}_p$ and has therefore a similar role as the controller index l in Chapters 2 and 3.

From a control perspective, it is essential that a globally asymptotically stabilizing PLQR can be determined for a p-periodic switching sequence. This motivates the following definitions:

Definition 4.9 *A p-periodic switching sequence $j(k)$ is called admissible if $(\overline{\boldsymbol{A}}_{0j}, \overline{\boldsymbol{B}}_{0j})$ is a stabilizable pair and $(\overline{\boldsymbol{C}}_{0j}, \overline{\boldsymbol{A}}_{0j})$ is a detectable pair.*

Definition 4.10 *The set of admissible p-periodic switching sequences is defined by*

$$\mathbb{J}_{p,\mathrm{adm}} = \{j(k) | j(k) = j(k+p) \ \forall k \in \mathbb{N}_0,\ j(k) \ \textit{admissible}\}. \tag{4.33}$$

The admissibility of a p-periodic switching sequence in view of Proposition 4.6 guarantees the existence of a globally asymptotically stabilizing PLQR. Note that w.l.o.g. $k_0 = 0$ is considered in Definition 4.9, cf. Remark 4.1.

Problem 4.2 can now be solved in two steps: First, a PLQR is computed for each admissible p-periodic switching sequence $j(k) \in \mathbb{J}_{p,\mathrm{adm}}$ (periodic control subproblem). Then, the optimal admissible p-periodic switching sequence $j^*(k) \in \mathbb{J}_{p,\mathrm{adm}}$ is searched exhaustively by comparing the resulting minimum cost $\boldsymbol{x}_0^T \boldsymbol{P}_j(0) \boldsymbol{x}_0$ for all $j(k) \in \mathbb{J}_{p,\mathrm{adm}}$ (periodic scheduling subproblem). This can be formalized as

Theorem 4.11 *Problem 4.2 is solved with minimum cost*

$$J_p^* = \min_{j(k)\in\mathbb{J}_{p,\mathrm{adm}}} \boldsymbol{x}_0^T \boldsymbol{P}_j(0)\boldsymbol{x}_0 \tag{4.34}$$

by the optimal p-periodic switching sequence

$$j^*(k) = \arg \min_{j(k)\in\mathbb{J}_{p,\mathrm{adm}}} \boldsymbol{x}_0^T \boldsymbol{P}_j(0)\boldsymbol{x}_0 \tag{4.35}$$

and the optimal p-periodic state feedback control law

$$\boldsymbol{u}^*(k) = -\boldsymbol{K}_{j^*}(k)\boldsymbol{x}(k) \tag{4.36}$$

where $\boldsymbol{P}_j(0)$ *and* $\boldsymbol{K}_{j^*}(k)$ *are given by (4.27) and (4.28).*

PROOF. The proof follows immediately from the discussion in Section 4.1.2. Note that w.l.o.g. $k_0 = 0$ is considered and $\boldsymbol{K}_{j^*}^*(k)$ is shortly written as $\boldsymbol{K}_{j^*}(k)$. □

An algorithm for solving Problem 4.2 based on Theorem 4.11 is given in Algorithm 4.1.

Algorithm 4.1 Offline Scheduling Algorithm

Input: $\boldsymbol{A}_{j(k)}$, $\boldsymbol{B}_{j(k)}$, $\boldsymbol{Q}_{1j(k)}$, $\boldsymbol{Q}_{2j(k)}$, $\boldsymbol{x}_0$, p

Output: Optimal p-periodic switching sequence $j^*(k)$, optimal PLQR $\boldsymbol{K}_{j^*}(k)$

1. Determine $\overline{\boldsymbol{A}}_{0j}$, $\overline{\boldsymbol{B}}_{0j}$, $\overline{\boldsymbol{C}}_{0j}$, $\overline{\boldsymbol{D}}_{0j}$ from (4.14) for each $j(k) \in \mathbb{J}_p$
2. Determine $\mathbb{J}_{p,\mathrm{adm}}$ by checking the stabilizability of $(\overline{\boldsymbol{A}}_{0j}, \overline{\boldsymbol{B}}_{0j})$ and the detectability of $(\overline{\boldsymbol{C}}_{0j}, \overline{\boldsymbol{A}}_{0j})$ for each $j(k) \in \mathbb{J}_p$
3. Determine $\overline{\boldsymbol{P}}_{0j}$ from the ARE (4.25) for each $j(k) \in \mathbb{J}_{p,\mathrm{adm}}$
4. Determine $\boldsymbol{P}_j(0)$ from (4.27) for each $j(k) \in \mathbb{J}_{p,\mathrm{adm}}$
5. Determine $j^*(k)$ from (4.35) by exhaustive search
6. Determine $\overline{\boldsymbol{A}}_{k_0j^*}$, $\overline{\boldsymbol{B}}_{k_0j^*}$, $\overline{\boldsymbol{C}}_{k_0j^*}$, $\overline{\boldsymbol{D}}_{k_0j^*}$ for $k_0 = 1, \ldots, p-1$ from (4.14)
7. Determine $\overline{\boldsymbol{K}}_{k_0j^*}$ for $k_0 = 0, \ldots, p-1$ from (4.24)
8. Determine $\boldsymbol{K}_{j^*}(k)$ from (4.28)

Remark 4.3. Algorithm 4.1 is performed completely offline and thus denoted as Offline Scheduling Algorithm. The obtained optimal p-periodic switching sequence $j^*(k)$ and optimal PLQR $\boldsymbol{K}_{j^*}(k)$ are referred to as periodic control and offline scheduling ($\mathrm{PCS}_{\mathrm{off}}$) strategy.

Remark 4.4. The optimal p-periodic switching sequence $j^*(k)$ obtained by Algorithm 4.1 depends on the initial state $\boldsymbol{x}_0$. For practical applications, an optimal switching sequence and control law for an arbitrary initial state is frequently required. In this case either the expected value of the cost

$$J_{p,\exp}^* = \min_{j(k)\in\mathbb{J}_{p,\mathrm{adm}}} \mathrm{E}\left(\boldsymbol{x}_0^T \boldsymbol{P}_j(0)\boldsymbol{x}_0\right) = \min_{j(k)\in\mathbb{J}_{p,\mathrm{adm}}} \mathrm{tr}\left(\boldsymbol{P}_j(0)\right) \tag{4.37}$$

assuming $\boldsymbol{x}_0$ as a Gaussian random variable with zero expected value and unit covariance matrix, see Lemma A.1, or the maximum value of the cost

$$J^*_{p,\max} = \min_{j(k)\in\mathbb{J}_{p,\mathrm{adm}}} \max_{\boldsymbol{x}_0\in S_{\mathrm{semi}}} \boldsymbol{x}_0^T \boldsymbol{P}_j(0)\boldsymbol{x}_0 = \min_{j(k)\in\mathbb{J}_{p,\mathrm{adm}}} \lambda_{\max}\left(\boldsymbol{P}_j(0)\right) \tag{4.38}$$

on the unit semi-hypersphere, see Lemmas A.2 and 2.3, can be used in (4.35) and step 5 of Algorithm 4.1. The expected value of the cost $J^*_{p,\mathrm{exp}}$ relates to the mean performance, the maximum value of the cost $J^*_{p,\max}$ to the worst-case performance.

Remark 4.5. The complexity of Algorithm 4.1 is principally determined by $|\mathbb{J}_p| = M^p$ and therefore grows exponentially with the period p. In steps 3 to 5 only the set of admissible p-periodic switching sequences with $|\mathbb{J}_{p,\mathrm{adm}}| \leq |\mathbb{J}_p|$ is considered whereby the complexity may reduce to some extent.

Remark 4.6. The period p can be considered as a design parameter. The choice of the period p can be based on the relations

$$|\mathbb{J}_{p_1}| < |\mathbb{J}_{p_2}| \quad \text{for } p_1 < p_2 \tag{4.39a}$$

$$\mathbb{J}_{p_1} \subset \mathbb{J}_{p_2} \quad \text{for } rp_1 = p_2,\ r = 2,3,\ldots \tag{4.39b}$$

$$\mathbb{J}_{p_1} \not\subset \mathbb{J}_{p_2} \quad \text{for } rp_1 \neq p_2,\ r = 2,3,\ldots. \tag{4.39c}$$

Relations (4.39a) and (4.39b) suggest that a larger period p leads to a larger number of variations and consequently to a lower cost. This may, however, not always be the case due to relation (4.39c). It is therefore reasonable to inspect several values for the period p.

Example 4.1 Consider the switched system (1.3) with the system and input matrices

$$\boldsymbol{A}_1 = \boldsymbol{A}_2 = \boldsymbol{A}_3 = \begin{pmatrix} 1.2 & 0 & 0 & 0 \\ 0 & 1.1 & 0 & 0 \\ 0 & 0 & 0.2 & 0 \\ 0 & 0 & 0 & 0.9 \end{pmatrix},\ \boldsymbol{B}_1 = \begin{pmatrix} 1 \\ 0 \\ 0 \\ 0 \end{pmatrix},\ \boldsymbol{B}_2 = \begin{pmatrix} 0 \\ 1 \\ 1 \\ 0 \end{pmatrix},\ \boldsymbol{B}_3 = \begin{pmatrix} 0 \\ 0 \\ 0 \\ 1 \end{pmatrix} \tag{4.40}$$

and the cost function (4.2) with the switched weighting matrices

$$\begin{aligned} \boldsymbol{Q}_{11} &= \boldsymbol{Q}_{12} = \boldsymbol{Q}_{13} = \boldsymbol{I} \\ \boldsymbol{Q}_{21} &= \boldsymbol{Q}_{22} = \boldsymbol{Q}_{23} = 5, \end{aligned} \tag{4.41}$$

whereby Problem 4.2 is characterized. This switched system may represent a networked embedded control system (NECS) consisting of an unstable first-order plant P_1 with the eigenvalue $\lambda_1 = 1.2$, an unstable second-order plant P_2 with the eigenvalues $\lambda_{21} = 1.1$ and $\lambda_{22} = 0.2$ and a stable first-order plant P_3 with the eigenvalue $\lambda_3 = 0.9$. The switching index $j(k)$ determines via the input matrices $\boldsymbol{B}_{j(k)}$ which plant $\mathrm{P}_{j(k)}$ is controlled at time instant k. The other plants remain uncontrolled. Therefore, the subsystems $(\boldsymbol{A}_{j(k)}, \boldsymbol{B}_{j(k)})$ are not stabilizable which affects the admissibility. Indeed, only p-periodic switching sequences which contain all switching indices $j(k) \in \mathbb{M}$ are admissible for this switched system as will be shown in Proposition 8.3. The switched system surely describes an NECS very simplistically. A more detailed NECS model is introduced in Chapter 7.

Problem 4.2 is solved for different periods $p \in \{3, \ldots, 12\}$ using Algorithm 4.1 where both the expected cost $J^*_{p,\text{exp}}$ according to (4.37) and the maximum cost $J^*_{p,\text{max}}$ according to (4.38) are considered. The number of p-periodic switching sequences $|\mathbb{J}_p|$, the number of admissible p-periodic switching sequences $|\mathbb{J}_{p,\text{adm}}|$, their ratio $|\mathbb{J}_{p,\text{adm}}|/|\mathbb{J}_p|$, the optimal admissible p-periodic switching sequence $j^*(k) \in \mathbb{J}_{p,\text{adm}}$, the expected cost $J^*_{p,\text{exp}}$ and maximum cost $J^*_{p,\text{max}}$ obtained for $j^*(k)$ as well as the computation time T_{comp} resulting for Algorithm 4.1 are indicated in Table 4.1.

p	$\lvert\mathbb{J}_p\rvert$	$\lvert\mathbb{J}_{p,\text{adm}}\rvert$	$\frac{\lvert\mathbb{J}_{p,\text{adm}}\rvert}{\lvert\mathbb{J}_p\rvert}$	$j^*(k)$	$J^*_{p,\text{exp}}$	T_{comp}
3	27	6	0.22	$(1,2,3)$	18.9129	0.02 s
4	81	36	0.44	$(1,2,2,3)$	18.4664	0.06 s
5	243	150	0.62	$(1,2,1,2,3)$	18.2704	0.20 s
6	729	540	0.74	$(1,2,1,2,3,2)$	18.1939	0.70 s
7	2187	1806	0.83	$(1,2,1,2,3,2,3)$	18.2045	2.51 s
8	6561	5796	0.88	$(1,2,1,2,3,2,2,3)$	18.1907	8.96 s
9	19683	18150	0.92	$(1,2,1,2,3,2,1,2,3)$	18.1786	39.85 s
10	59049	55980	0.95	$(1,2,1,2,3,2,1,2,3,2)$	18.1744	3.72 min
11	177147	171006	0.97	$(1,2,1,2,3,2,1,2,3,2,3)$	18.1726	27.48 min
12	531441	519156	0.98	$(1,2,1,2,3,2,1,2,3,2,3,2)$	18.1729	4.16 h
p	$\lvert\mathbb{J}_p\rvert$	$\lvert\mathbb{J}_{p,\text{adm}}\rvert$	$\frac{\lvert\mathbb{J}_{p,\text{adm}}\rvert}{\lvert\mathbb{J}_p\rvert}$	$j^*(k)$	$J^*_{p,\text{max}}$	T_{comp}
3	27	6	0.22	$(1,2,3)$	7.82309	0.05 s
4	81	36	0.44	$(1,2,2,3)$	7.01854	0.06 s
5	243	150	0.62	$(1,2,2,1,3)$	6.67657	0.20 s
6	729	540	0.74	$(1,2,2,1,2,3)$	6.47470	0.71 s
7	2187	1806	0.83	$(1,2,2,1,2,1,3)$	6.39393	2.50 s
8	6561	5796	0.88	$(1,2,2,1,2,1,2,3)$	6.35977	8.93 s
9	19683	18150	0.92	$(1,2,2,1,2,1,2,1,3)$	6.32376	40.61 s
10	59049	55980	0.95	$(1,2,2,1,2,1,2,1,3,1)$	6.32788	3.86 min
11	177147	171006	0.97	$(1,2,2,1,2,1,2,1,1,2,3)$	6.31268	29.36 min
12	531441	519156	0.98	$(1,2,2,1,2,1,2,1,1,2,1,3)$	6.30898	4.26 h

Table 4.1: Results for the PCS_{off} strategy

We first concentrate on the expected cost $J^*_{p,\text{exp}}$. The expected cost principally decreases with increasing period p but may also increase, e.g. between $p = 6$ and $p = 7$, confirming Remark 4.6. The decrease becomes insignificant for $p \geq 9$. Further increasing p then does not provide any benefit. Essentially, $p = 6$ may be a reasonable choice. The expected cost e.g. for $p = 6$ varies between 18.1939 and 82.6748 among all $j(k) \in \mathbb{J}_{p,\text{adm}}$, indicating that some $j(k)$ lead to very poor performance. The ratio $|\mathbb{J}_{p,\text{adm}}|/|\mathbb{J}_p|$ increases with p which is generally the case for NECSs. The computation time is acceptable even for large p.

We now inspect the maximum cost $J^*_{p,\max}$. The results are evidently very similar to those obtained for the expected cost $J^*_{p,\exp}$ except that the optimal admissible p-periodic switching sequences $j^*(k)$ usually differ.

The optimal admissible p-periodic switching sequences $j^*(k)$ reflect the dynamics of the plants. The unstable plant P_1 which has the larger growth rate is controlled priorly, then the unstable plant P_2 and then the stable plant P_3. □

Periodic control and offline scheduling in summary leads to an optimal p-periodic state feedback control law (4.36) and an optimal p-periodic switching sequence (4.35). Global asymptotic stability of the switched system (1.3) under the PCS_{off} strategy (4.35)/(4.36) is unlike under the RHCS strategy (3.3) guaranteed inherently, essentially due to using an infinite horizon. The performance is, however, degraded by imposing periodicity since the degrees of freedom for the optimization are reduced thereby. Furthermore, the optimal p-periodic switching sequence (4.35) is not adapted under disturbances since Algorithm 4.1 is performed completely offline. This may further impair the performance. Therefore, a periodic control and online scheduling providing an additional degree of freedom for the optimization and allowing an adaptation under disturbances while preserving the stability is desirable.

Theorem 4.11 indeed intriguingly parallels Theorem 2.2. The concepts presented in Chapters 2 and 3 on finite- and receding-horizon control and scheduling can thus be exploited for a periodic control and online scheduling. This is the subject of the following section. Particularly, the explicit solution proposed in Section 3.2 is revised to extend the Offline Scheduling Algorithm 4.1 to an Online Scheduling Algorithm. Furthermore, the relaxation introduced in Section 2.4 is utilized for complexity reduction.

4.2 Online Scheduling

4.2.1 Problem Formulation

Consider the discrete-time quadratic cost function

$$J_p(k) = \sum_{i=0}^{\infty} \left[\boldsymbol{x}^T(k+i)\boldsymbol{Q}_{1j(k+i)}\boldsymbol{x}(k+i) + \boldsymbol{u}^T(k+i)\boldsymbol{Q}_{2j(k+i)}\boldsymbol{u}(k+i) \right] \tag{4.42}$$

where $j(k)$ is a p-periodic switching sequence, $\boldsymbol{Q}_{1j(k)} \succeq \boldsymbol{0}$ and $\boldsymbol{Q}_{2j(k)} \succ \boldsymbol{0}$ are symmetric switched weighting matrices and the prediction horizon is infinite. Furthermore, k denotes the current time instant, i.e. the absolute time, and i denotes the time instant within the infinite prediction horizon, i.e. the relative time, see Section 1.4.4 for an illustration.

Problem 4.12 *For the switched system (1.3) and the current state $\boldsymbol{x}(k)$ find a control sequence $\boldsymbol{u}^*(k),\dots,\boldsymbol{u}^*(\infty)$ and a p-periodic switching sequence $j^*(k),\dots,j^*(\infty)$ such that the cost function (4.42) is minimized over the infinite prediction horizon, i.e.*

$$\begin{aligned}&\min_{\substack{\boldsymbol{u}(k),\dots,\boldsymbol{u}(\infty)\\ j(k),\dots,j(\infty)}} J_p(k)\\ &\text{subject to } \boldsymbol{x}(k+1+i)=\boldsymbol{A}_{j(k+i)}\boldsymbol{x}(k+i)+\boldsymbol{B}_{j(k+i)}\boldsymbol{u}(k+i),\ i=0,\dots,\infty.\end{aligned} \tag{4.43}$$

If the first element $\boldsymbol{u}^*(k)$ of the control sequence and the first element $j^*(k)$ of the p-periodic switching sequence are then applied to the switched system at the current time instant k, a periodic control and online scheduling (PCS_{on}) strategy is obtained.

4.2.2 Explicit Solution

Problem 4.12 implies solving Problem 4.2 at each time instant k for the current state $\boldsymbol{x}(k)$. The problem itself does not change over the time instants k, only the current state $\boldsymbol{x}(k)$ changes. The solution can therefore be determined offline w.l.o.g. for time instant $k=0$ and adapted online at each time instant k for the current state $\boldsymbol{x}(k)$. This can be formalized as

Theorem 4.13 *The solution to Problem 4.12 is given by the piecewise linear (PWL) state feedback control law*

$$\boldsymbol{u}^*(k)=-\boldsymbol{K}_v(0)\boldsymbol{x}(k) \quad \text{for} \quad \boldsymbol{x}(k)\in\mathcal{X}_v \tag{4.44}$$

where the regions $\mathcal{X}_v$ with $\bigcup_{j(i)\in\mathbb{J}_{p,\text{adm}}}\mathcal{X}_j=\mathbb{R}^{\text{n}}$ are described by

$$\mathcal{X}_v=\left\{\boldsymbol{x}(k)\mid \boldsymbol{x}^T(k)\boldsymbol{P}_v(0)\boldsymbol{x}(k)\le\boldsymbol{x}^T(k)\boldsymbol{P}_j(0)\boldsymbol{x}(k)\ \forall j(i);v(i),j(i)\in\mathbb{J}_{p,\text{adm}}\right\}. \tag{4.45}$$

Proof. The optimal switching index for the current state $\boldsymbol{x}(k)$ is given by

$$v(0)=j^*(0)=\arg\min_{j(i)\in\mathbb{J}_{p,\text{adm}}}\boldsymbol{x}^T(k)\boldsymbol{P}_j(0)\boldsymbol{x}(k), \tag{4.46}$$

utilizing Theorem 4.11 and in particular (4.35). This corresponds to

$$\boldsymbol{x}^T(k)\boldsymbol{P}_v(0)\boldsymbol{x}(k)\le\boldsymbol{x}^T(k)\boldsymbol{P}_j(0)\boldsymbol{x}(k) \quad \forall j(i)\in\mathbb{J}_{p,\text{adm}} \tag{4.47}$$

which leads to the regions (4.45). Note that $j(i)\in\mathbb{J}_{p,\text{adm}}$, $\boldsymbol{P}_j(i)$ and $\boldsymbol{K}_j(i)$ are w.l.o.g. related to $k=0$. □

Remark 4.7. The concept behind the PCS_{on} strategy (4.44) deserves some interpretation:

A region membership test based on (4.45) is performed first. The region membership test returns the region $\mathcal{X}_v$ associated to the p-periodic switching sequence $v(i)\in\mathbb{J}_{p,\text{adm}}$ yielding the minimum cost for the current state $\boldsymbol{x}(k)$. The first element $v(0)$ of this p-periodic switching sequence and the first element $\boldsymbol{K}_v(0)$ of the p-periodic feedback matrices associated to this p-periodic switching sequence are then used. Repeating this procedure at each time instant $k\in\mathbb{N}_0$ constitutes the PCS_{on} strategy (4.44).

Remark 4.8. The statements on the explicitness and well-posedness of the solution given in Remarks 3.3 and 3.4 and on the non-convexity of the regions given in Remark 3.5 analogously apply to Theorem 4.13.

Remark 4.9. Substituting (4.44) in (1.3) yields the autonomous PWL closed-loop system

$$\boldsymbol{x}(k+1) = \widetilde{\boldsymbol{A}}_v \boldsymbol{x}(k) \quad \text{for} \quad \boldsymbol{x}(k) \in \mathcal{X}_v, \quad \boldsymbol{x}(0) = \boldsymbol{x}_0 \tag{4.48}$$

with $\widetilde{\boldsymbol{A}}_v = \boldsymbol{A}_{v(0)} - \boldsymbol{B}_{v(0)} \boldsymbol{K}_v(0)$.

An algorithm for solving Problem 4.12 and implementing the PCS_{on} strategy (4.44) divides into an offline part (Algorithm 4.2) and an online part (Algorithms 4.3 and 4.4).

Algorithm 4.2 Online Scheduling Algorithm (Offline Part)

Input: $\boldsymbol{A}_{j(k)}$, $\boldsymbol{B}_{j(k)}$, $\boldsymbol{Q}_{1j(k)}$, $\boldsymbol{Q}_{2j(k)}$, p
Output: Switching index $j(0)$, DPRE sol. $\boldsymbol{P}_j(0)$, PLQR $\boldsymbol{K}_j(0)$ for each $j(i) \in \mathbb{J}_{p,\text{adm}}$

1. Determine $\overline{\boldsymbol{A}}_{0j}$, $\overline{\boldsymbol{B}}_{0j}$, $\overline{\boldsymbol{C}}_{0j}$, $\overline{\boldsymbol{D}}_{0j}$ from (4.14) for each $j(i) \in \mathbb{J}_p$
2. Determine $\mathbb{J}_{p,\text{adm}}$ by checking the stabilizability of $(\overline{\boldsymbol{A}}_{0j}, \overline{\boldsymbol{B}}_{0j})$ and the detectability of $(\overline{\boldsymbol{C}}_{0j}, \overline{\boldsymbol{A}}_{0j})$ for each $j(i) \in \mathbb{J}_p$
3. Determine $\overline{\boldsymbol{P}}_{0j}$ from the ARE (4.25) for each $j(i) \in \mathbb{J}_{p,\text{adm}}$
4. Determine $\boldsymbol{P}_j(0)$ from (4.27) for each $j(i) \in \mathbb{J}_{p,\text{adm}}$
5. Determine $\overline{\boldsymbol{K}}_{0j}$ from (4.24) for each $j(i) \in \mathbb{J}_{p,\text{adm}}$
6. Determine $\boldsymbol{K}_j(0)$ from (4.28) for each $j(i) \in \mathbb{J}_{p,\text{adm}}$
7. Store $j(0)$, $\boldsymbol{P}_j(0)$ and $\boldsymbol{K}_j(0)$ for each $j(i) \in \mathbb{J}_{p,\text{adm}}$

Algorithm 4.3 Online Scheduling Algorithm (Online Part)

Input: Switching index $j(0)$, DPRE sol. $\boldsymbol{P}_j(0)$, PLQR $\boldsymbol{K}_j(0)$ for each $j(i) \in \mathbb{J}_{p,\text{adm}}$
for each time instant k **do**
Measure the current state $\boldsymbol{x}(k)$
Determine the optimal switching index $v(0)$ from Algorithm 4.4
Determine the optimal control vector $\boldsymbol{u}^*(k) = \boldsymbol{K}_v(0)\boldsymbol{x}(k)$
Apply $v(0)$ and $\boldsymbol{u}^*(k)$ to the switched system (1.3)
end for

Algorithm 4.4 Region Membership Test

Input: Switching index $j(0)$, DPRE solution $\boldsymbol{P}_j(0)$ for each $j(i) \in \mathbb{J}_{p,\text{adm}}$; state $\boldsymbol{x}(k)$
Output: Optimal switching index $v(0)$ and therewith optimal switching sequence $v(i)$
$J_p^*(k) = \infty$ // *initialize cost*
for each $j(i) \in \mathbb{J}_{p,\text{adm}}$ **do**
if $\boldsymbol{x}^T(k)\boldsymbol{P}_j(0)\boldsymbol{x}(k) < J_p^*(k)$ **then**
$J_p^*(k) = \boldsymbol{x}^T(k)\boldsymbol{P}_j(0)\boldsymbol{x}(k)$ // *update cost*
$v(0) = j(0)$ // *update optimal switching index*
end if
end for

Remark 4.10. The complexity of Algorithms 4.2, 4.3 and 4.4 is generally characterized by $|\mathbb{J}_p| = M^p$ and thus grows exponentially with the period p. In Algorithms 4.3 and 4.4 and in steps 3 to 7 of Algorithm 4.2 only the set of admissible p-periodic switching sequences with $|\mathbb{J}_{p,\text{adm}}| \leq |\mathbb{J}_p|$ is considered whereby complexity may reduce. The online complexity, which is primarily determined by the Region Membership Test in Algorithm 4.4, may yet be prohibitive in practical applications, particularly for switched systems with fast dynamics. In this case, a complexity reduction is indispensable. Methods for reducing the online complexity are presented in Section 4.2.4.

Remark 4.11. The PCS_on strategy (4.44) is essentially an RHCS strategy relying on the PCS_off strategy (4.35)/(4.36). The switching sequence $j^*(k)$ resulting under the PCS_on strategy is usually aperiodic. The periodicity is only imposed as a "trick" to solve Problem 4.12 with an infinite horizon for the offline part. For the online part the receding horizon principle is utilized. The periodicity is overridden thereby and an additional degree of freedom for the optimization is gained. Therefore, the cost resulting from the PCS_on strategy is always less than or equal to the cost resulting from the PCS_off strategy as will be shown in Corollary 4.15.

Example 4.2 Reconsider the setup in Example 4.1 for the period $p = 6$. The switched system (4.40) under the PCS_off strategy (4.35)/(4.36) and under the PCS_on strategy (4.44) is simulated for the initial state $\tilde{\boldsymbol{x}}_0 = \begin{pmatrix} 1 & 5 & 5 & 1 \end{pmatrix}^T$ and for Gaussian random initial states with zero expected value and unit covariance matrix. The latter relates to a Monte Carlo simulation: If simulations are performed repeatedly for Gaussian random initial states, then the arithmetic mean cost $\overline{J}_p$ over all simulations converges to the expected cost $J^*_{p,\text{exp}}$ given by (4.37) due to the law of large numbers, see e.g. [PP02, Section 7.4] for details. For the given setup, 7500 simulations have proved sufficient for convergence. The mean cost $\overline{J}_p$ may also be interpreted as the mean cost resulting under random impulsive disturbances and thus also allows to conclude on the adaptivity of the PCS_on strategy under disturbances. Note that the PCS_off strategy has been designed for the initial state $\tilde{\boldsymbol{x}}_0$, the expected cost $J^*_{p,\text{exp}}$ and the maximum cost $J^*_{p,\max}$. The switching sequence $j^*(k)$, the cost J_p according to (4.2) resulting for the initial state $\tilde{\boldsymbol{x}}_0$ and the mean cost $\overline{J}_p$ resulting for random initial states are indicated in Table 4.2.

Strategy	Initial State	$j^*(k)$	J_p
PCS_off (designed for $\tilde{\boldsymbol{x}}_0$)	$\tilde{\boldsymbol{x}}_0$	$(2,2,1,2,1,3)$	149.303
PCS_off (designed for $J^*_{p,\text{exp}}$)	$\tilde{\boldsymbol{x}}_0$	$(1,2,1,2,3,2)$	209.034
PCS_off (designed for $J^*_{p,\max}$)	$\tilde{\boldsymbol{x}}_0$	$(1,2,2,1,2,3)$	193.882
PCS_on	$\tilde{\boldsymbol{x}}_0$	$(2,2,1,2,2,1,2,1,2,3,2,1,\ldots)$	147.987
Strategy	Initial State	$j^*(k)$	$\overline{J}_p$
PCS_off (designed for $J^*_{p,\text{exp}}$)	random	$(1,2,1,2,3,2)$	17.9436
PCS_off (designed for $J^*_{p,\max}$)	random	$(1,2,2,1,2,3)$	18.0405
PCS_on	random	adapted	15.5462

Table 4.2: Comparison of the PCS_off and PCS_on strategy ($p = 6$)

The $\mathrm{PCS_{on}}$ strategy leads to the smallest cost for both the initial state $\tilde{\boldsymbol{x}}_0$ and random initial states. Particularly for random initial states, the mean costs resulting under the $\mathrm{PCS_{off}}$ and $\mathrm{PCS_{on}}$ strategy differ significantly, which is attributed to the adaptivity of the $\mathrm{PCS_{on}}$ strategy. The switching sequences $j^*(k)$ resulting under the $\mathrm{PCS_{on}}$ strategy are aperiodic, confirming Remark 4.11. The online complexity for the $\mathrm{PCS_{on}}$ strategy may however be prohibitive since the region membership test comprises $|\mathbb{J}_{p,\mathrm{adm}}| = 540$ regions, cf. Table 4.1.

For the initial state $\tilde{\boldsymbol{x}}_0 = \begin{pmatrix}1 & 5 & 5 & 1\end{pmatrix}^T$, plant P_2 should be controlled priorly and more frequently. This is reflected by the switching sequences $j^*(k)$ resulting under the $\mathrm{PCS_{off}}$ strategy designed for $\tilde{\boldsymbol{x}}_0$ and the $\mathrm{PCS_{on}}$ strategy. However, the cost resulting for the $\mathrm{PCS_{off}}$ strategy designed for $\tilde{\boldsymbol{x}}_0$ is slightly larger due to the periodicity and admissibility constraints. The costs resulting for the $\mathrm{PCS_{off}}$ strategy designed for expected cost $J^*_{p,\mathrm{exp}}$ and maximum cost $J^*_{p,\mathrm{max}}$ are considerably larger since the initial state $\tilde{\boldsymbol{x}}_0$ is not regarded. The computation time resulting for Algorithm 4.2 corresponds approximately to the computation time resulting for Algorithm 4.1 which is given in Table 4.1. □

In the following section the stability of the switched system (1.3) under the $\mathrm{PCS_{on}}$ strategy (4.44) is proved by comparison to the $\mathrm{PCS_{off}}$ strategy (4.35)/(4.36).

4.2.3 Stability Analysis

Theorem 4.14 *The switched system (1.3) under the PCS_{on} strategy (4.44) or equivalently the PWL closed-loop system (4.48) is globally asymptotically stable.*

PROOF. Let $\boldsymbol{x}_{\mathrm{off}}(k)$ and $\boldsymbol{u}_{\mathrm{off}}(k)$ be the state and control trajectory of the switched system (1.3) under an arbitrary admissible p-periodic switching sequence $j(k) \in \mathbb{J}_{p,\mathrm{adm}}$ and the corresponding PLQR. This strategy may be interpreted as a suboptimal form of the $\mathrm{PCS_{off}}$ strategy (4.35)/(4.36). The cost function associated to the trajectory between time instant $k = k_{\mathrm{s}}$ and time instant $k = k_{\mathrm{f}}$ starting from state $\boldsymbol{x}(k_{\mathrm{s}})$ is defined as

$$J_{p,\mathrm{off}}(\boldsymbol{x}(k_{\mathrm{s}}), k_{\mathrm{s}}, k_{\mathrm{f}}, j_{k_{\mathrm{s}}}(i)) = \sum_{k=k_{\mathrm{s}}}^{k_{\mathrm{f}}} \left[\boldsymbol{x}^T_{\mathrm{off}}(k)\boldsymbol{Q}_{1j_{k_{\mathrm{s}}}(k-k_{\mathrm{s}})}\boldsymbol{x}_{\mathrm{off}}(k) + \boldsymbol{u}^T_{\mathrm{off}}(k)\boldsymbol{Q}_{2j_{k_{\mathrm{s}}}(k-k_{\mathrm{s}})}\boldsymbol{u}_{\mathrm{off}}(k)\right]. \tag{4.49}$$

The notation $j_{k_{\mathrm{s}}}(i)$ indicates that the p-periodic switching sequence $j(i)$ has been fixed at time instant k_{s}, e.g. by optimization. This notation will be frequently used within the remaining chapter and should therefore be borne in mind.

Furthermore, let $\boldsymbol{x}_{\mathrm{on}}(k)$ and $\boldsymbol{u}_{\mathrm{on}}(k)$ be the state and control trajectory of the switched system (1.3) under the $\mathrm{PCS_{on}}$ strategy (4.44). The cost function related to this trajectory between time instant $k = k_{\mathrm{s}}$ and time instant $k = k_{\mathrm{f}}$ starting from state $\boldsymbol{x}(k_{\mathrm{s}})$ is defined as

$$J_{p,\mathrm{on}}(\boldsymbol{x}(k_\mathrm{s}), k_\mathrm{s}, k_\mathrm{f}) = \sum_{k=k_\mathrm{s}}^{k_\mathrm{f}} \left[\boldsymbol{x}_\mathrm{on}^T(k)\boldsymbol{Q}_{1j_k^*(0)}\boldsymbol{x}_\mathrm{on}(k) + \boldsymbol{u}_\mathrm{on}^T(k)\boldsymbol{Q}_{2j_k^*(0)}\boldsymbol{u}_\mathrm{on}(k)\right] \tag{4.50}$$

where the optimal switching index $j_k^*(0)$ at time instant k results from

$$\begin{aligned} j_k^*(0) &= \arg \min_{j_k(i)\in\mathbb{J}_{p,\mathrm{adm}}} J_{p,\mathrm{off}}(\boldsymbol{x}_\mathrm{on}(k), k, \infty, j_k(i)) \\ &= \arg \min_{j_k(i)\in\mathbb{J}_{p,\mathrm{adm}}} \boldsymbol{x}_\mathrm{on}^T(k)\boldsymbol{P}_{j_k}(0)\boldsymbol{x}_\mathrm{on}(k) \end{aligned} \tag{4.51}$$

which follows directly by comparing (4.46) and (4.35) for a shifted initial time.

The cost function $J_{p,\mathrm{off}}(\boldsymbol{x}_0, 0, \infty, j_0(i))$ is finite for all admissible p-periodic switching sequences $j_0(i) \in \mathbb{J}_{p,\mathrm{adm}}$ as a direct consequence of Definition 4.9 and Proposition 4.6. Therefore, to prove global asymptotic stability of the switched system (1.3) under the $\mathrm{PCS_{on}}$ strategy (4.44) it must be shown that

$$J_{p,\mathrm{on}}(\boldsymbol{x}_0, 0, \infty) \leq J_{p,\mathrm{off}}(\boldsymbol{x}_0, 0, \infty, j_0(i)) \quad \forall j_0(i) \in \mathbb{J}_{p,\mathrm{adm}}. \tag{4.52}$$

Consider the cost function

$$J_{p,\mathrm{on\text{-}off}}(\boldsymbol{x}_0, r) = J_{p,\mathrm{on}}(\boldsymbol{x}_0, 0, r-1) + J_{p,\mathrm{off}}(\boldsymbol{x}_\mathrm{on}(r), r, \infty, j_r^*(i)) \tag{4.53}$$

which results when applying the $\mathrm{PCS_{on}}$ strategy between $k = 0$ and $k = r$ and then the $\mathrm{PCS_{off}}$ strategy between $k = r + 1$ and $k = \infty$. Clearly,

$$J_{p,\mathrm{on}}(\boldsymbol{x}_0, 0, \infty) = \lim_{r\to\infty} J_{p,\mathrm{on\text{-}off}}(\boldsymbol{x}_0, r). \tag{4.54}$$

Therefore, to prove (4.52) it is sufficient to show that

$$J_{p,\mathrm{on\text{-}off}}(\boldsymbol{x}_0, r) \leq J_{p,\mathrm{off}}(\boldsymbol{x}_0, 0, \infty, j_0(i)) \quad \forall j_0(i) \in \mathbb{J}_{p,\mathrm{adm}} \quad \forall r \in \mathbb{N}_0. \tag{4.55}$$

This proof follows by induction.

Induction Basis. At time instant $r = 0$, the optimal switching index under the $\mathrm{PCS_{on}}$ strategy is chosen as

$$j_0^*(0) = \arg \min_{j_0(i)\in\mathbb{J}_{p,\mathrm{adm}}} J_{p,\mathrm{off}}(\boldsymbol{x}_0, 0, \infty, j_0(i)). \tag{4.56}$$

Regarding that by definition $J_{p,\mathrm{on\text{-}off}}(\boldsymbol{x}_0, 0) = J_{p,\mathrm{off}}(\boldsymbol{x}_0, 0, \infty, j_0^*(i))$ yields

$$J_{p,\mathrm{on\text{-}off}}(\boldsymbol{x}_0, 0) \leq J_{p,\mathrm{off}}(\boldsymbol{x}_0, 0, \infty, j_0(i)) \quad \forall j_0(i) \in \mathbb{J}_{p,\mathrm{adm}}. \tag{4.57}$$

Thus, (4.55) is satisfied for $r = 0$.

Induction Step. Assume that (4.55) is satisfied for an arbitrary time instant $r \geq 0$, i.e.

$$J_{p,\mathrm{on\text{-}off}}(\boldsymbol{x}_0, r) \leq J_{p,\mathrm{off}}(\boldsymbol{x}_0, 0, \infty, j_0(i)) \quad \forall j_0(i) \in \mathbb{J}_{p,\mathrm{adm}} \quad (\textit{Induction Hypothesis}). \tag{4.58}$$

It must then be shown that (4.55) is also satisfied for time instant $r+1$. At time instant $r+1$, the optimal switching index under the $\mathrm{PCS_{on}}$ strategy is chosen as

$$j^*_{r+1}(0) = \arg \min_{j_{r+1}(i) \in \mathbb{J}_{p,\mathrm{adm}}} J_{p,\mathrm{off}}(\boldsymbol{x}_{\mathrm{on}}(r+1), r+1, \infty, j_{r+1}(i)). \tag{4.59}$$

Therefore,

$$J_{p,\mathrm{off}}(\boldsymbol{x}_{\mathrm{on}}(r+1), r+1, \infty, j^*_{r+1}(i)) \leq J_{p,\mathrm{off}}(\boldsymbol{x}_{\mathrm{on}}(r+1), r+1, \infty, j^*_r(i)) \tag{4.60}$$

with $j^*_{r+1}(i)$ indicating that a reoptimization is performed at time instant $r+1$ and $j^*_r(i)$ indicating that no reoptimization is performed but the p-periodic switching sequence fixed at time instant r is continued at time instant $r+1$. This principle is illustrated in Figure 4.3.

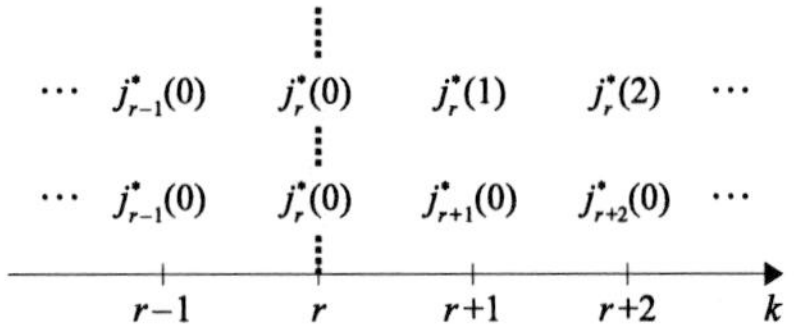

Figure 4.3: Reoptimization at each time instant k (bottom) versus reoptimization at time instants $k = 0, \ldots, r-1, r$ and continuation of the p-periodic switching sequence $j^*_r(k-r)$ at time instants $k = r+1, r+2, \ldots$ (top)

The fulfillment of (4.60) is ensured inherently since always $j^*_{r+1}(i) = j^*_r(i)$ can be chosen.

Adding $J_{p,\mathrm{on}}(\boldsymbol{x}_0, 0, r)$ on both sides of (4.60) and using (4.53) leads to

$$J_{p,\text{on-off}}(\boldsymbol{x}_0, r+1) \leq J_{p,\text{on-off}}(\boldsymbol{x}_0, r). \tag{4.61}$$

From the induction hypothesis (4.58) and inequality (4.61) it can be concluded that (4.55) is also satisfied for $r+1$. This completes the proof. □

Remark 4.12. Note that a similar technique has been utilized in [BÇH06, Theorem 2] for proving global asymptotic stability of a networked control system under the OPP strategy which is outlined in Section 4.2.4. This technique has been generalized here for proving global asymptotic stability of the switched system (1.3) under the $\mathrm{PCS_{on}}$ strategy (4.44).

The proof of Theorem 4.14 reveals an intriguing property of the $\mathrm{PCS_{on}}$ strategy:

Corollary 4.15 *The cost resulting from the PCS_{on} strategy (4.44) is always less than or equal to the cost resulting from the PCS_{off} strategy (4.35)/(4.36), i.e.*

$$J_{p,\mathrm{on}}(\boldsymbol{x}_0, 0, \infty) \leq J_{p,\mathrm{off}}(\boldsymbol{x}_0, 0, \infty, j_0(i)) \quad \forall j_0(i) \in \mathbb{J}_{p,\mathrm{adm}}. \tag{4.62}$$

PROOF. Follows immediately from (4.52). □

The proof of Theorem 4.14 furthermore facilitates a suboptimal PCS_{on} strategy which ensures global asymptotic stability:

Corollary 4.16 *The switched system (1.3) under the PCS_{on} strategy (4.44) or equivalently the PWL closed-loop system (4.48) remains globally asymptotically stable if instead of the optimal switching index $v(0) = j_k^*(0)$ according to (4.46)/(4.51) some suboptimal switching index $v(0) = j_k(0)$ fulfilling*

$$\boldsymbol{x}^T(k)\boldsymbol{P}_{j_k}(0)\boldsymbol{x}(k) \leq \boldsymbol{x}^T(k)\boldsymbol{P}_{j_{k-1}}(1)\boldsymbol{x}(k) \tag{4.63}$$

for the current state $\boldsymbol{x}(k)$ is applied at the current time instant k, where the suboptimal switching index $v(0) = j_0(0)$ at the initial time instant 0 can be chosen arbitrarily. If the suboptimal switching index at the initial time instant 0 is chosen as

$$v(0) = j_0^*(0) = \arg \min_{j_0(i) \in \mathbb{J}_{p,\text{adm}}} J_{p,\text{off}}(\boldsymbol{x}_0, 0, \infty, j_0(i)) = \arg \min_{j_0(i) \in \mathbb{J}_{p,\text{adm}}} \boldsymbol{x}_0 \boldsymbol{P}_{j_0}(0)\boldsymbol{x}_0, \tag{4.64}$$

then the cost resulting from this suboptimal PCS_{on} strategy furthermore remains less than or equal to the cost resulting from the PCS_{off} strategy (4.35)/(4.36).

PROOF. It can be easily seen that the proof of Theorem 4.14 remains valid if

- inequalities (4.52), (4.55) and (4.57) are fulfilled for some arbitrary switching sequence $j_0(i)$ instead of all switching sequences $j_0(i) \in \mathbb{J}_{p,\text{adm}}$,
- equation (4.53) is formulated for some suboptimal switching sequence $j_r(i)$ instead of the optimal switching sequence $j_r^*(i)$,
- inequality (4.60) is fulfilled for some suboptimal switching sequences $j_r(i)$ and $j_{r+1}(i)$ instead of the optimal switching sequences $j_r^*(i)$ and $j_{r+1}^*(i)$.

Fulfillment of inequality (4.60) must, however, be enforced separately. This is achieved via inequality (4.63) which with a change of variables is equivalent to inequality (4.60).

For getting smaller or equal cost than under the PCS_{off} strategy (4.35)/(4.36), however, inequalities (4.52) and (4.57) must be fulfilled for the optimal switching sequence $j_0^*(i)$ resulting under the PCS_{off} strategy (4.35)/(4.36). This is ensured by (4.64). Inequality (4.63) then enforces that $J_{p,\text{on-off}}(\boldsymbol{x}_0, r)$ is monotonically non-increasing. □

Remark 4.13. Corollary 4.16 deserves some interpretation:

The idea basically consists in determining a switching sequence $j_k(i)$ at the current time instant k for which a smaller or equal cost than for continuing with the switching sequence $j_{k-1}(i)$ determined at the preceding time instant $k-1$ is obtained. Applying then the switching index $j_k(0)$ and the feedback matrix $\boldsymbol{K}_{j_k}(0)$ at the current time instant k leads to a suboptimal PCS_{on} strategy. Global asymptotic stability of the switched system under this suboptimal PCS_{on} strategy is preserved. Starting at time instant 0 with the optimal switching sequence $j_0^*(i)$ resulting under the PCS_{off} strategy furthermore ensures that the cost resulting under the suboptimal PCS_{on} strategy remains smaller or equal than under the PCS_{off} strategy. Very loosely speaking: Find something better or equal than just continuing periodically and ideally start off well.

Corollary 4.16 can manifestly be utilized for reducing the online complexity. To this end, strategies for determining the suboptimal switching index $v(0) = j_k(0)$ fulfilling (4.63) with reduced and predefined complexity must be conceived. Such suboptimal PCS_{on} strategies are presented in Section 4.2.4.

Remark 4.14. Instead of the minimum cost for a given initial state $\boldsymbol{x}_0$, also the expected cost $J^*_{p,\text{exp}}$ according to (4.37) or the maximum cost $J^*_{p,\text{max}}$ according to (4.38) can be considered in (4.64). The cost relation to the PCS_{off} strategy formulated in Corollary 4.16 then holds w.r.t. $J^*_{p,\text{exp}}$ or $J^*_{p,\text{max}}$.

4.2.4 Complexity Reduction

Complexity Reduction based on Relaxation

The online complexity can be reduced employing the concept of relaxation. To this end, Problem 4.12 is generalized to

Problem 4.17 *For the switched system (1.3) and the current state $\boldsymbol{x}(k)$ find a control sequence $\boldsymbol{u}(k), \ldots, \boldsymbol{u}(\infty)$ and a p-periodic switching sequence $j(k), \ldots, j(\infty)$ such that the resulting cost $J_p(k)$ over the infinite prediction horizon is bounded by*

$$J_p^*(k) \leq J_p(k) \leq \alpha J_p^*(k) \tag{4.65}$$

where $J_p^*(k) = \min_{j(i) \in \mathbb{J}_{p,\text{adm}}} \boldsymbol{x}^T(k) \boldsymbol{P}_j(0) \boldsymbol{x}(k)$ is the minimum cost related to Problem 4.12 and $\alpha \in \mathbb{R}$ with $\alpha \geq 1$ is a relaxation factor.

If the first element $\boldsymbol{u}(k)$ of the control sequence and the first element $j(k)$ of the p-periodic switching sequence are then applied to the switched system at the current time instant k, a relaxed periodic control and online scheduling (RPCS_{on}) strategy is obtained.

Problem 4.17 can be approached in two steps: First, Problem 4.12 (offline part) is solved using Algorithm 4.2. Particularly, the DPRE solution $\boldsymbol{P}_j(0)$ for each $j(i) \in \mathbb{J}_{p,\text{adm}}$ is determined. Then, relaxed pruning is applied to determine a relaxed set of admissible p-periodic switching sequences $\hat{\mathbb{J}}_{p,\text{adm}} \subseteq \mathbb{J}_{p,\text{adm}}$. The explicit solution to Problem 4.17 finally follows from Theorem 4.13 by using $\hat{\mathbb{J}}_{p,\text{adm}}$ instead of $\mathbb{J}_{p,\text{adm}}$.

The relaxed pruning is based on

Theorem 4.18 (Relaxation Criterion) *If for each $\alpha \boldsymbol{P}_v(0)$ with $v(i) \in \mathbb{J}_{p,\text{adm}}$ and each*

$\boldsymbol{x}(k) \in \mathbb{R}^{\mathrm{n}}$ *there exists at least one* $\boldsymbol{P}_{\hat{j}}(0)$ *with* $\hat{j}(i) \in \hat{\mathbb{J}}_{p,\mathrm{adm}} \subseteq \mathbb{J}_{p,\mathrm{adm}}$ *such that*

$$\boldsymbol{x}^T(k)\alpha\boldsymbol{P}_v(0)\boldsymbol{x}(k) \geq \boldsymbol{x}^T(k)\boldsymbol{P}_{\hat{j}}(0)\boldsymbol{x}(k), \tag{4.66}$$

then $J_p(k) = \min_{\hat{j}(i)\in\hat{\mathbb{J}}_{p,\mathrm{adm}}} \boldsymbol{x}^T(k)\boldsymbol{P}_{\hat{j}}(0)\boldsymbol{x}(k)$ *satisfies (4.65).*

PROOF. Inequality (4.65) can be equivalently written as

$$\min_{j(i)\in\mathbb{J}_{p,\mathrm{adm}}} \boldsymbol{x}^T(k)\boldsymbol{P}_j(0)\boldsymbol{x}(k) \leq \min_{\hat{j}(i)\in\hat{\mathbb{J}}_{p,\mathrm{adm}}} \boldsymbol{x}^T(k)\boldsymbol{P}_{\hat{j}}(0)\boldsymbol{x}(k) \leq \min_{j(i)\in\mathbb{J}_{p,\mathrm{adm}}} \boldsymbol{x}^T(k)\alpha\boldsymbol{P}_j(0)\boldsymbol{x}(k). \tag{4.67}$$

Fulfillment of the first inequality in (4.67) is ensured since $\hat{\mathbb{J}}_{p,\mathrm{adm}} \subseteq \mathbb{J}_{p,\mathrm{adm}}$, fulfillment of the second inequality in (4.67) due to (4.66). □

Remark 4.15. The relaxation criterion can in general not be evaluated exactly since the switching sequence $\hat{j}(i) \in \hat{\mathbb{J}}_{p,\mathrm{adm}}$ fulfilling (4.66) may vary over $\boldsymbol{x}(k) \in \mathbb{R}^{\mathrm{n}}$. Instead, a common switching sequence $\hat{j}(i) = w(i) = \mathrm{const.}$ with $w(i) \in \hat{\mathbb{J}}_{p,\mathrm{adm}} \setminus \{v(i)\}$ can be considered for all states $\boldsymbol{x}(k) \in \mathbb{R}^{\mathrm{n}}$ analogously to Theorem 2.5. Alternatively, a convex combination of $\boldsymbol{P}_{\hat{j}}(0)$ can be utilized deploying the S-procedure analogously to Theorem 2.6. The latter approach is less conservative but computationally more demanding as pointed out in Remarks 2.7 and 2.8. For conciseness only the latter approach will be considered in the following.

Remark 4.16. For $\alpha = 1$ the minimum cost $J_p^*(k)$ is preserved. The online complexity may still be reduced since the DPRE solutions $\boldsymbol{P}_v(0)$ not contributing to the minimum cost $J_p^*(k)$, i.e. for which

$$J_p^*(k) = \min_{j(i)\in\mathbb{J}_{p,\mathrm{adm}}} \boldsymbol{x}^T(k)\boldsymbol{P}_j(0)\boldsymbol{x}(k) = \min_{j(i)\in\mathbb{J}_{p,\mathrm{adm}}\setminus\{v(i)\}} \boldsymbol{x}(k)^T\boldsymbol{P}_j(0)\boldsymbol{x}(k) \tag{4.68}$$

holds, are pruned.

Remark 4.17. A relaxed DPRE solution $\alpha\boldsymbol{P}_j(0)$ is always dominated by the associated DPRE solution $\boldsymbol{P}_j(0)$, i.e. $\boldsymbol{P}_j(0) \preceq \alpha\boldsymbol{P}_j(0)$ for all $j(i) \in \mathbb{J}_{p,\mathrm{adm}}$. Therefore, (4.66) is satisfied trivially for $\hat{\mathbb{J}}_{p,\mathrm{adm}} = \mathbb{J}_{p,\mathrm{adm}}$. In terms of complexity reduction it is rather required to fulfill (4.66) with a minimum number of DPRE solutions $|\hat{\mathbb{J}}_{p,\mathrm{adm}}|$.

An efficient domination criterion based on the S-procedure can be formulated as

Theorem 4.19 (Domination Criterion) *If there exist non-negative scalars* $\xi_{\hat{j}}$ *and* $\boldsymbol{P}_{\hat{j}}(0)$ *with* $\hat{j}(i) \in \hat{\mathbb{J}}_{p,\mathrm{adm}}$ *such that*

$$\sum_{\hat{j}(i)\in\hat{\mathbb{J}}_{p,\mathrm{adm}}} \xi_{\hat{j}} = 1 \quad \textit{and} \quad \alpha\boldsymbol{P}_v(0) \succeq \sum_{\hat{j}(i)\in\hat{\mathbb{J}}_{p,\mathrm{adm}}} \xi_{\hat{j}}\boldsymbol{P}_{\hat{j}}(0), \tag{4.69}$$

then $\alpha\boldsymbol{P}_v(0)$ *with* $v(i) \in \mathbb{J}_{p,\mathrm{adm}}$ *is dominated.*

PROOF. Follows the same lines as the proof of Theorem 2.6. □

Algorithm 4.5 Relaxed Pruning Algorithm

Input: DPRE solutions $\boldsymbol{P}_j(0), j(i) \in \mathbb{J}_{p,\mathrm{adm}}$ before relaxed pruning, relaxation factor α
Output: DPRE solutions $\boldsymbol{P}_{\hat{j}}(0), \hat{j}(i) \in \hat{\mathbb{J}}_{p,\mathrm{adm}}$ after relaxed pruning

$\hat{\mathbb{J}}_{p,\mathrm{adm}} = \{\arg\min_{j(i) \in \mathbb{J}_{p,\mathrm{adm}}} \mathrm{tr}(\boldsymbol{P}_j(0))\}$ // *initialize relaxed set (e.g. based on trace)*
for each $v(i) \in \mathbb{J}_{p,\mathrm{adm}}$ **do**
  if $\nexists\, \xi_{\hat{j}} \geq 0,\ \hat{j}(i) \in \hat{\mathbb{J}}_{p,\mathrm{adm}}$: $\sum_{\hat{j}(i) \in \hat{\mathbb{J}}_{p,\mathrm{adm}}} \xi_{\hat{j}} = 1,\ \alpha \boldsymbol{P}_v(0) \succeq \sum_{\hat{j}(i) \in \hat{\mathbb{J}}_{p,\mathrm{adm}}} \xi_{\hat{j}} \boldsymbol{P}_{\hat{j}}(0)$ **then**
    $\hat{\mathbb{J}}_{p,\mathrm{adm}} = \hat{\mathbb{J}}_{p,\mathrm{adm}} \cup \{v(i)\}$ // *add switching sequence*
  end if
end for

An algorithm for constructing the relaxed set of admissible p-periodic switching sequences $\hat{\mathbb{J}}_{p,\mathrm{adm}}$ based on Theorem 4.19 is given in Algorithm 4.5.

The algorithm is based on adding switching sequences $j(i)$ from the set of admissible p-periodic switching sequences $\mathbb{J}_{p,\mathrm{adm}}$ to the relaxed set of admissible p-periodic switching sequences $\hat{\mathbb{J}}_{p,\mathrm{adm}}$ until all relaxed DPRE solutions $\alpha\boldsymbol{P}_j(0)$, $j(i) \in \mathbb{J}_{p,\mathrm{adm}}$ are dominated. For complexity reduction it is crucial to first add switching sequences $j(i)$ with DPRE solutions $\alpha\boldsymbol{P}_j(0)$ that are presumably not dominated. This can be achieved by presorting the DPRE solutions by trace, i.e. w.r.t. the expected value of the cost, see Remark 4.4.

Global asymptotic stability of the switched system (1.3) under the $\mathrm{RPCS_{on}}$ strategy (4.44) is guaranteed under a mild condition. For formulating this condition, the cyclic shift operator must be introduced:

Definition 4.20 *The cyclic shift operator* $S : \mathbb{J}_p \to \mathbb{J}_p$ *is defined by*

$$S(j(i)) = j((i+1) \bmod p). \tag{4.70}$$

Furthermore, applying the cyclic shift operator $i_\mathrm{s} \in \mathbb{N}_0$ *times is described by*

$$S^{i_\mathrm{s}}(j(i)) = j((i+i_\mathrm{s}) \bmod p). \tag{4.71}$$

Example 4.3 Consider the 5-periodic switching sequence $j(i) = (1,2,1,2,3)$. Then $S(j(i)) = (2,1,2,3,1)$, $S^2(j(i)) = (1,2,3,1,2)$, $S^3(j(i)) = (2,3,1,2,1)$ and so on. □

Remark 4.18. The cyclic shift operator S is strongly related to the initial time i_0. Applying $S^{i_\mathrm{s}}(j(i))$ corresponds to cyclically increasing i_0 of $j(i)$ by i_s. This relation together with Remark 4.1 and Definition 4.9 entails that admissibility is preserved under the cyclic shift operator, i.e. if $j(i) \in \mathbb{J}_{p,\mathrm{adm}}$, then also $S^{i_\mathrm{s}}(j(i)) \in \mathbb{J}_{p,\mathrm{adm}}$ for all $i_\mathrm{s} \in \mathbb{N}_0$. This is a subtle point in the forthcoming Theorem 4.21.

Theorem 4.21 *The switched system (1.3) under the* RPCS_{on} *strategy (4.44) or equivalently the PWL closed-loop system (4.48) is globally asymptotically stable for* $\alpha = 1$ *in general and for* $\alpha > 1$ *if for each* $\boldsymbol{P}_v(0)$ *with* $v(i) \in \hat{\mathbb{J}}_{p,\mathrm{adm}}$ *and each* $\boldsymbol{x}(k) \in \mathbb{R}^\mathrm{n}$ *there exists at least one* $\boldsymbol{P}_{\hat{j}}(0)$ *with* $\hat{j}(i) \in \hat{\mathbb{J}}_{p,\mathrm{adm}}$ *such that*

$$\boldsymbol{x}^T(k)\boldsymbol{P}_{v_\mathrm{s}}(0)\boldsymbol{x}(k) \geq \boldsymbol{x}^T(k)\boldsymbol{P}_{\hat{j}}(0)\boldsymbol{x}(k) \tag{4.72}$$

with $v_s(i) = S(v(i)) \in \mathbb{J}_{p,\text{adm}}$.

PROOF. The proof follows the same lines as the proof of Theorem 4.14 when considering $\hat{j}(i) \in \hat{\mathbb{J}}_{p,\text{adm}}$ instead of $j(i) \in \mathbb{J}_{p,\text{adm}}$. However, the fulfillment of (4.60) is not ensured inherently due to pruning and must therefore be proved separately.

For $\alpha = 1$, fulfillment of (4.60) follows from the relaxation criterion in Theorem 4.18. Inequality (4.60) can be equivalently written as

$$\boldsymbol{x}_{\text{on}}^T(r+1)\boldsymbol{P}_{\hat{j}_{r+1}^*}(0)\boldsymbol{x}_{\text{on}}(r+1) \leq \boldsymbol{x}_{\text{on}}^T(r+1)\boldsymbol{P}_{\hat{j}_r^*}(1)\boldsymbol{x}_{\text{on}}(r+1) \tag{4.73}$$

with $\boldsymbol{P}_{\hat{j}_{r+1}^*}(0)$ characterizing the cost if the current p-periodic switching sequence $j_{r+1}^*(i)$ is determined by reoptimization at time instant $r+1$ and $\boldsymbol{P}_{\hat{j}_r^*}(1)$ characterizing the cost if the preceding p-periodic switching sequence $j_r^*(i)$ fixed at time instant r is continued at time instant $r+1$, recall Figure 4.3 for an illustration of this principle. The relaxation criterion guarantees that for each $\hat{j}_r^*(i) \in \mathbb{J}_{p,\text{adm}}$ and each $\boldsymbol{x}_{\text{on}}(r+1) \in \mathbb{R}^n$ there exists at least one $\hat{j}_{r+1}^*(i) \in \hat{\mathbb{J}}_{p,\text{adm}}$ such that (4.73) and therefore also (4.60) is fulfilled. Note that this even holds if $\boldsymbol{P}_{\hat{j}_r^*}(1)$ has been pruned, i.e. if $\hat{j}_r^*(i) \in \mathbb{J}_{p,\text{adm}}$ but $\hat{j}_r^*(i) \notin \hat{\mathbb{J}}_{p,\text{adm}}$.

For $\alpha > 1$, the relaxation criterion only guarantees that for each $\hat{j}_r^*(i) \in \mathbb{J}_{p,\text{adm}}$ and each $\boldsymbol{x}_{\text{on}}(r+1) \in \mathbb{R}^n$ there exists at least one $\hat{j}_{r+1}^*(i) \in \hat{\mathbb{J}}_{p,\text{adm}}$ such that

$$\boldsymbol{x}_{\text{on}}^T(r+1)\boldsymbol{P}_{\hat{j}_{r+1}^*}(0)\boldsymbol{x}_{\text{on}}(r+1) \leq \alpha\boldsymbol{x}_{\text{on}}^T(r+1)\boldsymbol{P}_{\hat{j}_r^*}(1)\boldsymbol{x}_{\text{on}}(r+1) \tag{4.74}$$

is fulfilled. Therefore, if $\boldsymbol{P}_{\hat{j}_r^*}(1)$ has been pruned, i.e. $\hat{j}_r^*(i) \notin \hat{\mathbb{J}}_{p,\text{adm}}$, (4.73) may not be satisfied. Consequently, (4.73) must be checked separately. To formulate an appropriate condition, (4.73) should be abstracted from $j_r^*(i)$ and $j_{r+1}^*(i)$. The cyclic shift operator can employed to this end. Specifically, (4.73) can be abstracted to (4.72) by changing the variables and using $\boldsymbol{P}_v(1) = \boldsymbol{P}_{v_s}(0)$. □

Remark 4.19. Theorem 4.21 deserves some interpretation for $\alpha > 1$:

Global asymptotic stability of the switched system under the RPCS$_{\text{on}}$ strategy is guaranteed for $\alpha > 1$ if for each sequence $v(i)$ contained in $\hat{\mathbb{J}}_{p,\text{adm}}$ there exists a sequence $\hat{j}(i)$ contained in $\hat{\mathbb{J}}_{p,\text{adm}}$ which leads to smaller or equal cost than the cyclically shifted sequence $v_s(i) = S(v(i))$ contained in $\mathbb{J}_{p,\text{adm}}$. Notably, this condition is fulfilled trivially if the cyclically shifted sequence $v_s(i)$ is contained in $\hat{\mathbb{J}}_{p,\text{adm}}$ since then $\hat{j}(i) = v_s(i)$ can be chosen. The condition must thus only be evaluated if the cyclically shifted sequence $v_s(i)$ is not contained in $\hat{\mathbb{J}}_{p,\text{adm}}$.

Remark 4.20. The stability condition in Theorem 4.21 for $\alpha > 1$ can not be checked exactly but by utilizing a common switching index or the S-procedure, see Remark 4.15.

Remark 4.21. Global asymptotic stability of the switched system under the RPCS$_{\text{on}}$ strategy can always be guaranteed for $\alpha > 1$ by iteratively adding those cyclically shifted p-periodic switching sequences $v_s(i)$ to the relaxed set of admissible p-periodic

Algorithm 4.6 Stability Enforcement Algorithm

Input: $\boldsymbol{P}_{j(0)}$, $j(i) \in \mathbb{J}_{p,\mathrm{adm}}$ and $\boldsymbol{P}_{\hat{j}(0)}$, $\hat{j}(i) \in \hat{\mathbb{J}}_{p,\mathrm{adm}}$ before stability enforcement
Output: $\boldsymbol{P}_{\hat{j}(0)}$, $\hat{j}(i) \in \hat{\mathbb{J}}_{p,\mathrm{adm}}$ after stability enforcement
  for each $v(i) \in \hat{\mathbb{J}}_{p,\mathrm{adm}}$ **do**
    if $v_\mathrm{s}(i) = S(v(i)) \notin \hat{\mathbb{J}}_{p,\mathrm{adm}}$ **then**
      if $\nexists\, \xi_{\hat{j}} \geq 0, \hat{j}(i) \in \hat{\mathbb{J}}_{p,\mathrm{adm}}: \sum_{\hat{j}(i)\in\hat{\mathbb{J}}_{p,\mathrm{adm}}} \xi_{\hat{j}} = 1, \boldsymbol{P}_{v_\mathrm{s}}(0) \succeq \sum_{\hat{j}(i)\in\hat{\mathbb{J}}_{p,\mathrm{adm}}} \xi_{\hat{j}} \boldsymbol{P}_{\hat{j}}(0)$ **then**
        $\hat{\mathbb{J}}_{p,\mathrm{adm}} = \hat{\mathbb{J}}_{p,\mathrm{adm}} \cup \{v_\mathrm{s}(i)\}$ // *add cyclically shifted switching sequence*
      end if
    end if
  end for

switching sequences $\hat{\mathbb{J}}_{p,\mathrm{adm}}$ for which the stability condition given in Theorem 4.21 is not fulfilled. Based on this idea, Algorithm 4.6 is formulated for enforcing global asymptotic stability, where the stability condition given in Theorem 4.21 is evaluated utilizing the S-procedure. Note that if a cyclically shifted p-periodic switching sequence $v_\mathrm{s}(i)$ is added to the relaxed set of admissible p-periodic switching sequences $\hat{\mathbb{J}}_{p,\mathrm{adm}}$, then the associated cyclically shifted p-periodic switching sequence $S(v_\mathrm{s}(i))$ will be checked for fulfilling the stability condition in a subsequent iteration (for-loop over increasing $\hat{\mathbb{J}}_{p,\mathrm{adm}}$).

Corollary 4.22 *The cost resulting from an RPCS$_{on}$ strategy (4.44) fulfilling the stability condition in Theorem 4.21 and the cost resulting from the PCS$_{off}$ strategy (4.35)/(4.36) are related by*

$$J_{p,\mathrm{on}}(\boldsymbol{x}_0, 0, \infty) \leq \alpha J_{p,\mathrm{off}}(\boldsymbol{x}_0, 0, \infty, j_0(i)) \quad \forall j_0(i) \in \mathbb{J}_{p,\mathrm{adm}}. \tag{4.75}$$

If furthermore the relaxed set of admissible p-periodic switching sequences $\hat{\mathbb{J}}_{p,adm}$ contains the optimal admissible p-periodic switching sequence $j^*(i)$ resulting from the PCS$_{off}$ strategy (4.35)/(4.36), then

$$J_{p,\mathrm{on}}(\boldsymbol{x}_0, 0, \infty) \leq J_{p,\mathrm{off}}(\boldsymbol{x}_0, 0, \infty, j_0(i)) \quad \forall j_0(i) \in \mathbb{J}_{p,\mathrm{adm}}. \tag{4.76}$$

PROOF. We first prove the cost relation (4.75). For an RPCS$_\mathrm{on}$ strategy (4.44) fulfilling the stability condition in Theorem 4.21, inequality (4.52) corresponds to

$$J_{p,\mathrm{on}}(\boldsymbol{x}_0, 0, \infty) \leq J_{p,\mathrm{off}}(\boldsymbol{x}_0, 0, \infty, \hat{j}_0(i)) \quad \forall \hat{j}_0(i) \in \hat{\mathbb{J}}_{p,\mathrm{adm}}. \tag{4.77}$$

Furthermore, the relaxation criterion in Theorem 4.18 implies that

$$J_{p,\mathrm{off}}(\boldsymbol{x}_0, 0, \infty, \hat{j}_0(i)) \leq \alpha J_{p,\mathrm{off}}(\boldsymbol{x}_0, 0, \infty, j_0(i)) \quad \forall \hat{j}_0(i) \in \hat{\mathbb{J}}_{p,\mathrm{adm}} \quad \forall j_0(i) \in \mathbb{J}_{p,\mathrm{adm}}, \tag{4.78}$$

completing the proof.

We now verify the cost relation (4.76). For getting smaller or equal cost than the one under the PCS$_\mathrm{off}$ strategy (4.35)/(4.36), the optimal admissible p-periodic switching sequence $j^*(i)$ resulting from the $\mathrm{PCS_{off}}$ strategy (4.35)/(4.36) must be selectable at time instant $k = 0$, cf. inequality (4.57). Thus, $j^*(i) \in \hat{\mathbb{J}}_{p,\mathrm{adm}}$ is required. By (4.72) it is then guaranteed that $J_{p,\text{on-off}}(\boldsymbol{x}_0, r)$ is monotonically non-increasing. □

Remark 4.22. The cost relation (4.76) can always be enforced by adding the optimal admissible p-periodic switching sequence $j^*(i)$ resulting from the $\mathrm{PCS_{off}}$ strategy (4.35)/(4.36) to the relaxed set of admissible p-periodic switching sequences $\hat{\mathbb{J}}_{p,\mathrm{adm}}$ before evaluating the stability condition in Theorem 4.21 or the Stability Enforcement Algorithm 4.6. Note that $j^*(i)$ can be determined for a given initial state according to (4.35), the expected cost according to (4.37) or the maximum cost according to (4.38). The cost relation (4.76) then applies accordingly.

Corollary 4.23 *The switched system (1.3) under an $RPCS_{on}$ strategy (4.44) fulfilling the stability condition in Theorem 4.21 or equivalently the PWL closed-loop system (4.48) remains globally asymptotically stable if a suboptimal switching index $v(0) = \hat{j}_k(0)$ fulfilling*

$$\boldsymbol{x}^T(k)\boldsymbol{P}_{\hat{j}_k}(0)\boldsymbol{x}(k) \leq \boldsymbol{x}^T(k)\boldsymbol{P}_{\hat{j}_{k-1}}(1)\boldsymbol{x}(k) \tag{4.79}$$

for the current state $\boldsymbol{x}(k)$ is applied at the current time instant k, where the suboptimal switching index $v(0) = \hat{j}_0(0)$ at the initial time instant 0 can be chosen arbitrarily. If furthermore the suboptimal switching index at the initial time instant 0 is chosen as

$$v(0) = \hat{j}_0(0) = \arg \min_{\hat{j}_0(i) \in \hat{\mathbb{J}}_{p,\mathrm{adm}}} J_{p,\mathrm{off}}(\boldsymbol{x}_0, 0, \infty, \hat{j}_0(i)) = \arg \min_{\hat{j}_0(i) \in \hat{\mathbb{J}}_{p,\mathrm{adm}}} \boldsymbol{x}_0 \boldsymbol{P}_{\hat{j}_0}(0)\boldsymbol{x}_0, \tag{4.80}$$

then the cost resulting from this strategy and the cost resulting from the PCS_{off} strategy (4.35)/(4.36) remain related by (4.75) if the optimal sequence $j^*(i)$ resulting from the PCS_{off} strategy is not contained in $\hat{\mathbb{J}}_{p,\mathrm{adm}}$ and by (4.76) otherwise.

PROOF. Follows the same lines as the proof of Corollary 4.16. □

Remark 4.23. The statement on the cost relation to the $\mathrm{PCS_{off}}$ strategy in Remark 4.14 analogously applies to Corollary 4.23.

Example 4.4 Reconsider the setup in Example 4.1 for the period $p = 6$. Problem 4.17 is solved for different relaxation factors α using Algorithm 4.2 for determining the switching sequences, DPRE solutions and PLQRs, Algorithm 4.5 for relaxed pruning and Algorithm 4.6 for enforcing stability. Noteworthy, for all considered relaxation factors α the optimal admissible p-periodic switching sequence $j^*(i)$ resulting from the $\mathrm{PCS_{off}}$ strategy (4.35)/(4.36) (designed for the expected cost) has been contained in the relaxed set of admissible p-periodic switching sequences $\hat{\mathbb{J}}_{p,\mathrm{adm}}$ before stability enforcement (SE) and therefore had not to be added to enforce the cost relation (4.76). The results are given in Table 4.3, specifically

- the number of admissible p-periodic switching sequences $|\hat{\mathbb{J}}_{p,\mathrm{adm}}|$ after SE,
- the number of admissible p-periodic switching sequences $|\hat{\mathbb{J}}_{p,\mathrm{adm}}|$ before SE,

- the number of cyclically shifted admissible p-periodic switching sequences $v_s(i)$ that turned out to be already contained in $\hat{\mathbb{J}}_{p,\text{adm}}$ during SE,
- the number of cyclically shifted admissible p-periodic switching sequences $v_s(i)$ that turned out to be not contained in $\hat{\mathbb{J}}_{p,\text{adm}}$ during SE but had not to be added.

α	$\lvert\hat{\mathbb{J}}_{p,\text{adm}}\rvert$ after SE	$\lvert\hat{\mathbb{J}}_{p,\text{adm}}\rvert$ before SE	No. $v_s(i)$ contained	No. $v_s(i)$ not added	$\overline{J}_p$	T_{comp}
2.0000	6	1	1	0	15.9180	1.11 s
1.7500	12	2	2	0	15.9082	1.53 s
1.5000	18	3	3	0	15.8590	2.47 s
1.2500	42	7	7	0	15.5949	11.99 s
1.1000	85	19	17	2	15.5657	25.74 s
1.0750	132	29	21	8	15.5542	36.77 s
1.0500	149	41	34	7	15.5629	45.61 s
1.0250	213	90	69	21	15.5499	1.29 min
1.0100	304	217	174	43	15.5464	2.51 min
1.0075	327	263	212	51	15.5464	2.84 min
1.0050	353	311	262	49	15.5462	3.01 min
1.0025	391	365	315	50	15.5462	3.39 min
1.0010	417	397	350	47	15.5462	3.66 min
1.0000	460	460	420	40	15.5462	4.20 min

Table 4.3: Results for the RPCS_{on} strategy ($p = 6$)

Evidently, many $v_s(i)$ turn out to be already contained in $\hat{\mathbb{J}}_{p,\text{adm}}$ during SE, particularly for small α. Furthermore, often not all $v_s(i)$ not contained in $\hat{\mathbb{J}}_{p,\text{adm}}$ had to be added during SE, again particularly for small α. Consequently, $\lvert\hat{\mathbb{J}}_{p,\text{adm}}\rvert$ is not overly increased during SE. The number of admissible p-periodic switching sequences $\lvert\hat{\mathbb{J}}_{p,\text{adm}}\rvert$ after SE is significantly smaller than the number of admissible p-periodic switching sequences $\lvert\mathbb{J}_{p,\text{adm}}\rvert = 540$ resulting for the PCS_{on} strategy (Table 4.1). This applies also for small α. Noteworthy, $\lvert\hat{\mathbb{J}}_{p,\text{adm}}\rvert$ after SE is smaller than $\lvert\mathbb{J}_{p,\text{adm}}\rvert$ even for $\alpha = 1$, which confirms Remark 4.16. The online complexity can thus be reduced significantly while relaxing optimality only slightly.

The switched system (4.40) under the resulting RPCS_{on} strategies (4.44) is simulated for random initial states, see Example 4.2. The resulting mean cost $\overline{J}_p$ is given in Table 4.3. The mean cost generally decreases with the relaxation factor α but may also increase, e.g. between $\alpha = 1.075$ and $\alpha = 1.05$, which is attributed to the sufficiency of the domination criterion given in Theorem 4.19 and the Relaxation Algorithm 4.5. For $\alpha \leq 1.005$ the mean costs differ only marginally in lower decimal places (not indicated). The mean costs resulting under the RPCS_{on} strategy for $\alpha = 1$ and under the PCS_{on} strategy are as expected identical, cf. Table 4.2. A reasonable compromise between online complexity (determined by $\lvert\hat{\mathbb{J}}_{p,\text{adm}}\rvert$ after SE) and mean cost $\overline{J}_p$ may be given for $\alpha = 1.25$. □

Complexity Reduction based on Heuristics

Various heuristics for reducing the online complexity with guaranteed stability and cost can be constructed based on Corollary 4.16 as pointed out in Remark 4.13. One possible heuristic is proposed in the following.

The idea consists in assigning to each sequence $j(i) \in \mathbb{J}_{p,\text{adm}}$ a set $\mathbb{J}_{p,\text{adm}}(j(i)) \subset \mathbb{J}_{p,\text{adm}}$ which has a predefined cardinality $|\mathbb{J}_{p,\text{adm}}(j(i))| = M'$ and contains $j(i)$ itself and sequences with large, medium and small similarity to $j(i)$. By including sequences with different similarity, the diversity of the sequences is preserved. The similarity of two sequences $j(i)$ and $j'(i)$ can be measured in different ways, e.g. by the Hamming distance [Ham50] which is defined as

$$d\left(j(i), j'(i)\right) = \sum_{i=0}^{p-1} [j(i) \neq j'(i)] \tag{4.81}$$

and satisfies $0 \le d(j(i), j'(i)) \le p$ and $d(j(i), j(i)) = 0$. The Hamming distance $d\left(j(i), j'(i)\right)$ thus corresponds to the number of different elements of the two sequences $j(i)$ and $j'(i)$.

The set $\mathbb{J}_{p,\text{adm}}(j(i))$ to a sequence $j(i) \in \mathbb{J}_{p,\text{adm}}$ can be constructed as follows:

- The set $\mathbb{J}_{p,\text{adm}}$ is first partitioned into subsets $\mathbb{J}_{p,\text{adm}}(j(i), z)$ whose elements have the same Hamming distance z to $j(i)$, i.e.

$$\mathbb{J}_{p,\text{adm}} = \bigcup_{z=0}^{p} \mathbb{J}_{p,\text{adm}}(j(i), z) \tag{4.82}$$

  with $\mathbb{J}_{p,\text{adm}}(j(i), z) = \{j'(i) \,|j'(i) \in \mathbb{J}_{p,\text{adm}}, d(j(i), j'(i)) = z\}$.

- The set $\mathbb{J}_{p,\text{adm}}(j(i))$ is then built from subsets $\mathbb{J}_{p,\text{adm,red}}(j(i), z) \subseteq \mathbb{J}_{p,\text{adm}}(j(i), z)$, i.e.

$$\mathbb{J}_{p,\text{adm}}(j(i)) = \bigcup_{z=0}^{p} \mathbb{J}_{p,\text{adm,red}}(j(i), z). \tag{4.83}$$

  The subsets $\mathbb{J}_{p,\text{adm,red}}(j(i), z)$ result from randomly removing elements from the subsets $\mathbb{J}_{p,\text{adm}}(j(i), z)$ such that

$$\frac{|\mathbb{J}_{p,\text{adm,red}}(j(i), z)|}{|\mathbb{J}_{p,\text{adm}}(j(i))|} \approx \frac{|\mathbb{J}_{p,\text{adm}}(j(i), z)|}{|\mathbb{J}_{p,\text{adm}}|} \tag{4.84a}$$

$$j(i) \in \mathbb{J}_{p,\text{adm}}(j(i)) \tag{4.84b}$$

$$|\mathbb{J}_{p,\text{adm}}(j(i))| = M' \tag{4.84c}$$

  is fulfilled. The condition (4.84a) ensures that the distribution of the Hamming distances and therefore the similarity distribution is preserved. This can usually be achieved only approximately since the cardinalities are integers. The distribution of the Hamming distances and the similarity distribution may also be changed by including weighting factors in condition (4.84a). The condition (4.84b) guarantees fulfillment of the stability condition given in Corollary 4.16 as will become clear in the following. An illustration of the partitioning for $p = 3$ is given in Figure 4.4.

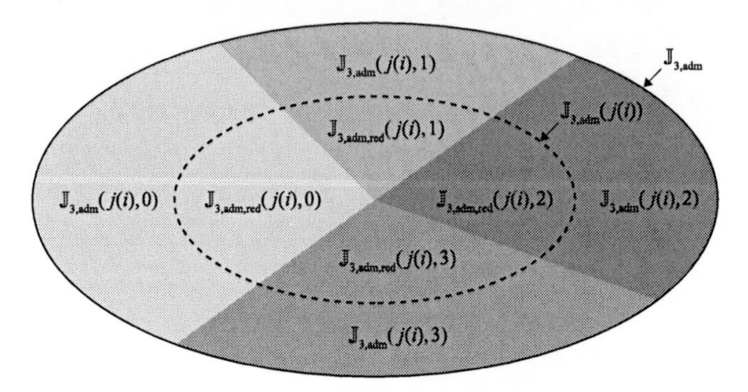

Figure 4.4: Illustration of the partitioning for $p = 3$

Example 4.5 Consider the set

$$\mathbb{J}_{3,\text{adm}} = \{(1,1,1),(1,1,2),(1,2,1),(1,2,2),(2,1,1),(2,1,2),(2,2,1),(2,2,2)\}.$$

Consider now the sequence $(1,2,1) \in \mathbb{J}_{3,\text{adm}}$, resulting in

$$\begin{aligned}
&\mathbb{J}_{3,\text{adm}}((1,2,1),0) = \{(1,2,1)\}, && |\mathbb{J}_{3,\text{adm}}((1,2,1),0)| = 1\\
&\mathbb{J}_{3,\text{adm}}((1,2,1),1) = \{(1,1,1),(1,2,2),(2,2,1)\}, && |\mathbb{J}_{3,\text{adm}}((1,2,1),1)| = 3\\
&\mathbb{J}_{3,\text{adm}}((1,2,1),2) = \{(1,1,2),(2,1,1),(2,2,2)\}, && |\mathbb{J}_{3,\text{adm}}((1,2,1),2)| = 3\\
&\mathbb{J}_{3,\text{adm}}((1,2,1),3) = \{(2,1,2)\}, && |\mathbb{J}_{3,\text{adm}}((1,2,1),3)| = 1.
\end{aligned}$$

One possible choice of the subsets $\mathbb{J}_{3,\text{adm,red}}((1,2,1),z)$ for a predefined $M' = 4$ then consists in

$$\begin{aligned}
&\mathbb{J}_{3,\text{adm,red}}((1,2,1),0) = \{(1,2,1)\}, && |\mathbb{J}_{3,\text{adm,red}}((1,2,1),0)| = 1\\
&\mathbb{J}_{3,\text{adm,red}}((1,2,1),1) = \{(1,1,1)\}, && |\mathbb{J}_{3,\text{adm,red}}((1,2,1),1)| = 1\\
&\mathbb{J}_{3,\text{adm,red}}((1,2,1),2) = \{(1,1,2)\}, && |\mathbb{J}_{3,\text{adm,red}}((1,2,1),2)| = 1\\
&\mathbb{J}_{3,\text{adm,red}}((1,2,1),3) = \{(2,1,2)\}, && |\mathbb{J}_{3,\text{adm,red}}((1,2,1),3)| = 1,
\end{aligned}$$

leading to

$$\mathbb{J}_{3,\text{adm}}((1,2,1)) = \{(1,1,1),(1,1,2),(1,2,1),(2,1,2)\}.$$

□

The PCS_{on} strategy (4.44) can be modified to a heuristic periodic control and online scheduling (HPCS_{on}) strategy based on the sets $\mathbb{J}_{p,\text{adm}}(j(i))$:

The sets $\mathbb{J}_{p,\text{adm}}(j(i))$ are computed offline for each sequence $j(i) \in \mathbb{J}_{p,\text{adm}}$ and stored. The switching index is determined online from

$$v(0) = j_k(0) = \arg \min_{j_k(i) \in \mathbb{J}_{p,\text{adm}}(S(j_{k-1}(i)))} \boldsymbol{x}^T(k) \boldsymbol{P}_{j_k}(0) \boldsymbol{x}(k) \tag{4.87}$$

where $j_0(0)$ can be chosen arbitrarily or according to (4.64). The region membership test at the current time instant k is thus restricted to the set $\mathbb{J}_{p,\text{adm}}(S(j_{k-1}(i)))$ associated to the cyclically shifted sequence $S(j_{k-1}(i))$ of the sequence $j_{k-1}(i)$ determined at the preceding time instant $k-1$. The online complexity is given by $|\mathbb{J}_{p,\text{adm}}(S(j_{k-1}(i)))| = M'$ and can thus be predefined via M'. The stability condition given in Corollary 4.16 is trivially fulfilled since the cyclically shifted sequence $S(j_{k-1}(i))$ is contained in the set $\mathbb{J}_{p,\text{adm}}(S(j_{k-1}(i)))$ due to condition (4.84b) and therefore $v(0) = S(j_{k-1}(0))$ is always selectable. If furthermore $j_0(0)$ is chosen according to (4.64), then also the cost relation to the PCS_{off} strategy formulated in Corollary 4.16 holds. However, the region membership test must then be performed for the complete set $\mathbb{J}_{p,\text{adm}}$ at time instant $k = 0$, which may be overly time-consuming. Alternatively, the expected cost $J_{p,\text{exp}}$ according to (4.37) or the maximum cost $J_{p,\max}$ according to (4.38) can be considered in (4.64) as outlined in Remark 4.14. The cost relation to the PCS_{off} strategy then applies accordingly. Algorithms 4.2, 4.3 and 4.4 can be easily modified for the HPCS_{on} strategy.

Example 4.6 Reconsider the setup in Example 4.1 for the period $p = 6$. The set $\mathbb{J}_{p,\text{adm}}(j(i))$ for each sequence $j(i) \in \mathbb{J}_{p,\text{adm}}$ is determined for $M' \in \{6, 42\}$. Since the procedure for determining the sets $\mathbb{J}_{p,\text{adm}}(j(i))$ is based on randomness, hundred sets have been determined for each $j(i) \in \mathbb{J}_{p,\text{adm}}$ and $M' \in \{6, 42\}$. The switched system (4.40) under the resulting HPCS_{on} strategies (4.87) is simulated for random initial states, see Example 4.2, where $j_0(0)$ is chosen according to (4.64) considering the initial state $\boldsymbol{x}_0$, the expected cost $J_{p,\text{exp}}$ or the maximum cost $J_{p,\max}$. The minimum, mean and maximum of the mean cost $\overline{J}_p$ over the hundred sets are listed in Table 4.4. Furthermore, the mean computation time $\overline{T}_{\text{comp}}$ over the hundred sets which results for performing the modified Algorithm 4.2 and determining the sets $\mathbb{J}_{p,\text{adm}}(j(i))$ is indicated.

M'	$j_0(0)$ chosen	$\overline{J}_p$ (minimum)	$\overline{J}_p$ (mean)	$\overline{J}_p$ (maximum)	$\overline{T}_{\text{comp}}$
6	based on $\boldsymbol{x}_0$	15.5805	15.5895	15.5999	2.45 s
6	based on $J^*_{p,\text{exp}}$	15.7878	16.2518	16.9407	2.45 s
6	based on $J^*_{p,\max}$	15.8890	16.3426	16.9596	2.45 s
42	based on $\boldsymbol{x}_0$	15.5603	15.5654	15.5722	4.09 s
42	based on $J^*_{p,\text{exp}}$	15.5956	15.6465	15.7358	4.09 s
42	based on $J^*_{p,\max}$	15.6045	15.6570	15.7927	4.09 s

Table 4.4: Results for the HPCS_{on} strategy ($p = 6$)

The mean cost varies considerably due to the randomness in determining the sets $\mathbb{J}_{p,\text{adm}}(j(i))$. Therefore, repeating the simulation for a large number of sets for each

$j(i) \in \mathbb{J}_{p,\mathrm{adm}}$ may be considered to select the favorable sets. This may, however, be very time-consuming.

The mean cost resulting for $j_0(0)$ chosen based on $\boldsymbol{x}_0$ is very small, also in comparison to the mean cost resulting under the RPCS$_{\mathrm{on}}$ strategy for the same online complexity $|\hat{\mathbb{J}}_{p,\mathrm{adm}}| = M' \in \{6, 42\}$, cf. Table 4.3. However, choosing $j_0(0)$ based on $\boldsymbol{x}_0$ may be very time-consuming. Furthermore, under disturbances $j_0(0)$ would actually have to be chosen persistently. In this regard, the results must be taken with care.

The mean of mean cost resulting for $j_0(0)$ chosen based on $J^*_{p,\mathrm{exp}}$ or $J^*_{p,\mathrm{max}}$ is slightly larger than the mean cost resulting under the RPCS$_{\mathrm{on}}$ strategy for the same online complexity for both $M' \in \{6, 42\}$. The minimum of the mean cost is, however, smaller for $M' = 6$. □

Remark 4.24. Similar considerations can be made based on Corollary 4.23, allowing to combine the RPCS$_{\mathrm{on}}$ and HPCS$_{\mathrm{on}}$ strategy.

Complexity Reduction based on Optimal Pointer Placement

Another approach for reducing the online complexity is proposed in [BÇH06]. The idea is to first determine the optimal admissible p-periodic switching sequence $j^*(i)$ resulting under the PCS$_{\mathrm{off}}$ strategy where the cost for a given initial state $\boldsymbol{x}_0$ based on (4.35), the expected cost $J^*_{p,\mathrm{exp}}$ based on (4.37) or the maximum cost $J^*_{p,\mathrm{max}}$ based on (4.38) can be considered. Then the cyclically shifted p-periodic switching sequences $S^{i_s}(j^*(i))$ and the associated feedback matrices $\boldsymbol{K}_{S^{i_s}(j^*)}(0)$ and DPRE solutions $\boldsymbol{P}_{S^{i_s}(j^*)}(0)$ for all $i_s \in \{0, \ldots, p-1\}$ are computed and stored. These steps are performed offline. The switching index $v(0)$ and feedback matrix $\boldsymbol{K}_v(0)$ are adjusted online at each time instant k by optimizing over i_s for the current state $\boldsymbol{x}(k)$, i.e.

$$v(0) = S^{i_s^*}(j^*(0)) = \arg \min_{i_s \in \{0,\ldots,p-1\}} \boldsymbol{x}^T(k) \boldsymbol{P}_{S^{i_s}(j^*)}(0) \boldsymbol{x}(k). \tag{4.88}$$

The online complexity is consequently determined by the period p. This procedure is called Optimal Pointer Placement (OPP) where i_s^* is denoted as optimal pointer. The OPP strategy (4.88) is essentially a specialization of the PCS$_{\mathrm{on}}$ strategy (4.44) since $\{S^{i_s}(j^*(i)), i_s \in \{0, \ldots, p-1\}\} \subseteq \mathbb{J}_{p,\mathrm{adm}}$. Algorithms 4.2, 4.3 and 4.4 can thus be easily modified for the OPP strategy.

The switched system (1.3) under the OPP strategy (4.88) is globally asymptotically stable. Furthermore, the cost resulting from the OPP strategy is always less than or equal to the cost resulting from the PCS$_{\mathrm{off}}$ strategy (4.35)/(4.36), i.e.

$$J_{p,\mathrm{OPP}}(\boldsymbol{x}_0, 0, \infty) \leq J_{p,\mathrm{off}}(\boldsymbol{x}_0, 0, \infty, j_0(i)) \quad \forall j(i) \in \mathbb{J}_{p,\mathrm{adm}} \tag{4.89}$$

where $J_{p,\mathrm{OPP}}(\boldsymbol{x}_0, 0, \infty)$ is defined analogously to (4.50) w.r.t. $\boldsymbol{x}_0$, $J^*_{p,\mathrm{exp}}$ or $J^*_{p,\mathrm{max}}$. A proof of these properties is given in [BÇH06, Theorem 2] and can also be inferred from Corollary 4.16.

Example 4.7 Reconsider the setup in Example 4.1 for the period $p = 6$. The switched system (4.40) under the OPP strategy (4.88) is simulated for the initial state $\tilde{\boldsymbol{x}}_0$ and for random initial states as specified in Example 4.2. Note that the OPP strategy has been designed for the initial state $\tilde{\boldsymbol{x}}_0$, the expected cost $J^*_{p,\text{exp}}$ or the maximum cost $J^*_{p,\max}$. The cost J_p resulting for the initial state $\tilde{\boldsymbol{x}}_0$ according to (4.2) and the mean cost $\overline{J}_p$ resulting for random initial states are indicated in Table 4.5.

Strategy	Initial State	$j(k)$	J_p
OPP (designed for $\tilde{\boldsymbol{x}}_0$)	$\tilde{\boldsymbol{x}}_0$	$(2,2,1,2,2,1,2,1,2,1,3,2,\ldots)$	148.409
OPP (designed for $J^*_{p,\text{exp}}$)	$\tilde{\boldsymbol{x}}_0$	$(2,2,1,2,1,2,1,2,3,2,1,2,\ldots)$	150.650
OPP (designed for $J^*_{p,\max}$)	$\tilde{\boldsymbol{x}}_0$	$(2,2,1,2,2,1,2,1,2,3,2,1,\ldots)$	148.267
Strategy	Initial State	$j(k)$	$\overline{J}_p$
OPP (designed for $\tilde{\boldsymbol{x}}_0$)	random	adapted	15.9322
OPP (designed for $J^*_{p,\text{exp}}$)	random	adapted	15.9180
OPP (designed for $J^*_{p,\max}$)	random	adapted	16.1373

Table 4.5: Results for the OPP strategy ($p = 6$)

For the initial state $\tilde{\boldsymbol{x}}_0$, interestingly the OPP strategy designed for $J^*_{p,\max}$ and not the one designed for $\tilde{\boldsymbol{x}}_0$ leads to the smallest cost. For random initial states, the OPP strategy designed for $J^*_{p,\text{exp}}$ yields the smallest mean cost. The mean costs resulting under the OPP strategy designed for $J^*_{p,\text{exp}}$ and under the RPCS$_{\text{on}}$ strategy designed for $\alpha = 2$, which have the same online complexity $|\hat{\mathbb{J}}_{p,\text{adm}}| = p = 6$, are identical, see Table 4.3. The reason is that $\hat{\mathbb{J}}_{p,\text{adm}}$ considered in the RPCS$_{\text{on}}$ strategy for $\alpha = 2$ consists of all cyclic shifts of $j^*(i)$ utilized in the OPP strategy designed for $J^*_{p,\text{exp}}$. The OPP strategy and the RPCS$_{\text{on}}$ strategy are then equivalent. □

Comparison of the RPCS$_{\text{on}}$, HPCS$_{\text{on}}$ and OPP Strategy

The RPCS$_{\text{on}}$, HPCS$_{\text{on}}$ and OPP strategy are compared in Table 4.6 in terms of their online complexity and their cost relation to the PCS$_{\text{off}}$ strategy and the PCS$_{\text{on}}$ strategy where $J_{p,\text{RPCS}_{\text{on}}}(\boldsymbol{x}_0, 0, \infty)$, $J_{p,\text{HPCS}_{\text{on}}}(\boldsymbol{x}_0, 0, \infty)$ and $J_{p,\text{OPP}}(\boldsymbol{x}_0, 0, \infty)$ are defined analogously to (4.50).

Strategy	Online Complexity	Cost Relations
RPCS$_{\text{on}}$	$\lvert\hat{\mathbb{J}}_{p,\text{adm}}\rvert^1 \leq M^p$	$J_{p,\text{on}}(\boldsymbol{x}_0,0,\infty) \leq J_{p,\text{RPCS}_{\text{on}}}(\boldsymbol{x}_0,0,\infty) \leq J_{p,\text{off}}(\boldsymbol{x}_0,0,\infty,j_0(i))^2$
HPCS$_{\text{on}}$	M'	$J_{p,\text{on}}(\boldsymbol{x}_0,0,\infty) \leq J_{p,\text{HPCS}_{\text{on}}}(\boldsymbol{x}_0,0,\infty) \leq J_{p,\text{off}}(\boldsymbol{x}_0,0,\infty,j_0(i))^3$
OPP	p	$J_{p,\text{on}}(\boldsymbol{x}_0,0,\infty) \leq J_{p,\text{OPP}}(\boldsymbol{x}_0,0,\infty) \leq J_{p,\text{off}}(\boldsymbol{x}_0,0,\infty,j_0(i))$

[1]adjustable via α [2]must be enforced, cf. Remark 4.22 [3]if $j_0(0)$ is chosen according to (4.64)

Table 4.6: Comparison of online complexity and cost relations

The online complexity can be influenced in all strategies: indirectly by adjusting the relaxation factor α in the RPCS$_{\text{on}}$ strategy, directly by predefining the cardinality M' in the HPCS$_{\text{on}}$ strategy and directly by selecting the period p in the OPP strategy. In the RPCS$_{\text{on}}$ and HPCS$_{\text{on}}$ strategy separate parameters α and M' are available for influencing the online complexity while in the OPP strategy the period p which is also a design parameter is utilized which surely limits the flexibility. The cost relations to the PCS$_{\text{off}}$ strategy and the PCS$_{\text{on}}$ strategy are identical for all strategies. For the RPCS$_{\text{on}}$ strategy, however, the cost relation must be enforced.

Overall, the RPCS$_{\text{on}}$ strategy may be the premier choice. The choice of a strategy ultimately depends on the application. Since all strategies are highly formalized, comparative simulations can be easily performed to facilitate the choice.

Example 4.8 The major results of Examples 4.2, 4.4, 4.6 and 4.7 are summarized in Table 4.7. Specifically, the online complexity characterized by the number of regions, the mean cost $\overline{J}_{p,\text{strat}}$ resulting for random initial states and the ratio $\overline{J}_{p,\text{strat}}/\overline{J}_{p,\text{PCS}_{\text{on}}}$ are indicated.

Strategy		Online Complexity	$\overline{J}_{p,\text{strat}}$	$\frac{\overline{J}_{p,\text{strat}}}{\overline{J}_{p,\text{PCS}_{\text{on}}}}$
PCS$_{\text{off}}$	(designed for $J^*_{p,\text{exp}}$)	negligible	17.9436	1.1542
PCS$_{\text{on}}$		$\lvert\hat{\mathbb{J}}_{p,\text{adm}}\rvert = 540$	15.5462	1.0000
RPCS$_{\text{on}}$	(designed for $\alpha = 2$)	$\lvert\hat{\mathbb{J}}_{p,\text{adm}}\rvert = 6$	15.9180	1.0239
HPCS$_{\text{on}}$	(designed for $J^*_{p,\text{exp}}$, minimum)	$M' = 6$	15.7878	1.0155
HPCS$_{\text{on}}$	(designed for $J^*_{p,\text{exp}}$, mean)	$M' = 6$	16.2518	1.0454
OPP	(designed for $J^*_{p,\text{exp}}$)	$p = 6$	15.9180	1.0239
RPCS$_{\text{on}}$	(designed for $\alpha = 1.25$)	$\lvert\hat{\mathbb{J}}_{p,\text{adm}}\rvert = 42$	15.5949	1.0031
HPCS$_{\text{on}}$	(designed for $J^*_{p,\text{exp}}$, minimum)	$M' = 42$	15.5956	1.0032
HPCS$_{\text{on}}$	(designed for $J^*_{p,\text{exp}}$, mean)	$M' = 42$	15.6465	1.0065

Table 4.7: Comparison of the PCS$_{\text{off}}$, PCS$_{\text{on}}$, RPCS$_{\text{on}}$, HPCS$_{\text{on}}$ and OPP strategy ($p = 6$)

The PCS$_{\text{off}}$ strategy leads to the largest mean cost; the online complexity is negligible since only the p-periodic switching sequence must be followed. The PCS$_{\text{on}}$ strategy yields the smallest mean cost; the online complexity, however, is considerable. The RPCS$_{\text{on}}$, HPCS$_{\text{on}}$ and OPP strategy allow for a compromise between complexity and performance. The mean costs are given in Table 4.7 for two numbers of regions:

For $\lvert\hat{\mathbb{J}}_{p,\text{adm}}\rvert = M' = p = 6$, the RPCS$_{\text{on}}$ and OPP strategy generally lead to smallest mean cost. The mean cost resulting under the HPCS$_{\text{on}}$ strategy strongly depends on the selection of the utilized sets. On the one hand, very small mean costs can be achieved if favorable sets are determined by simulation. On the other hand, this procedure is very time-consuming.

For $\lvert\hat{\mathbb{J}}_{p,\text{adm}}\rvert = M' = 42$, the RPCS$_{\text{on}}$ strategy leads to the smallest mean cost. Moreover, for $\lvert\hat{\mathbb{J}}_{p,\text{adm}}\rvert = M' = 42$ significantly smaller mean costs than for $\lvert\hat{\mathbb{J}}_{p,\text{adm}}\rvert = M' = 6$

are obtained. Indeed, the mean costs are very close to the mean cost resulting under the $\mathrm{PCS}_{\mathrm{on}}$ strategy. Notably, an OPP strategy with this number of regions can not be determined within reasonable computation time. □

5 Conclusions and Future Work

5.1 Conclusions

A comprehensive framework for optimal control and scheduling of switched systems has been presented. An overview of this framework is given in Figure 5.1.

	Receding-Horizon Control and Scheduling	**Periodic Control and Scheduling**
Optimization	**Finite-Horizon Control and Scheduling** • ***Dynamic Programming*** (Sec. 2.2, Thm. 2.2) • ***Dynamic Programming with Pruning*** (Sec. 2.3) • ***Relaxed Dynamic Programming*** (Sec. 2.4)	• ***Periodic Control and Exhaustive Search*** (Sec. 4.1.3, Thm. 4.11, Alg. 4.1)
Complexity Reduction (Offline Part)	• ***Pruning*** ✓ (Thms. 2.4, 2.5, 2.6, Algs. 2.1, 2.2) • ***Relaxation*** ✓ (Thms. 2.8, 2.9, Alg. 2.3, adj. via α ✓)	• ***none*** ×
Complexity Reduction (Online Part)	• ***not required***	• ***PCS_{on}*** × (Algs. 4.2, 4.3, 4.4) • ***$RPCS_{on}$*** ✓ (Thms. 4.18, 4.19, Alg. 4.5, adj. via α ✓) • ***$HPCS_{on}$*** ✓ (Cor. 4.16, adjustable via M' ✓) • ***OPP*** ✓ (Cor. 4.16, not adjustable ×)
Explicit Solution	• ***PWL State Feedback Control Law*** ✓(Thm. 3.2)	• ***PWL State Feedback Control Law*** ✓(Thm. 4.13)
Stability	• ***not guaranteed inherently*** × — RRHCS / non-stab. • ***A Priori Stability Conditon*** (Terminal Weighting Matrix) (Thms. 3.3, 3.4) — RRHCS: × ; non-stab.: × • ***A Posteriori Stability Criteria*** (PWQ Lyapunov Functions) (Thms. 3.6, 3.8, Lem. 3.7) — RRHCS: ✓ ; non-stab.: × • ***A Posteriori Stability Criterion*** (Value Fcn. as Lyapunov Fcn.) (Thm. 3.9) — RRHCS: ✓ ; non-stab.: ✓	• ***guaranteed inherently*** ✓ • ***PCS_{off}*** (Prop. 4.6, Def. 4.9) • ***PCS_{on}*** (Thm. 4.14) • ***$RPCS_{on}$***(Thm. 4.21, Alg. 4.6, under mild conditions) • ***$HPCS_{on}$***(Cor. 4.16) • ***OPP*** (Cor. 4.16)
Adaptivity	• ***RHCS, RRHCS*** ✓	• ***PCS_{off}*** × • ***PCS_{on}, $RPCS_{on}$, $HPCS_{on}$, OPP*** ✓

Figure 5.1: Framework for optimal control and scheduling of switched systems

Figure 5.1 also provides a comparison of receding-horizon control and scheduling and periodic control and scheduling in terms of optimization, complexity reduction for the offline part and online part, explicit solution, stability and adaptivity under disturbances. Furthermore, advantages ✓ and disadvantages ✗ are indicated.

The main advantage of receding-horizon control and scheduling consists in a complexity reduction during optimization based on pruning and relaxation. The main disadvantage consists in the lack of an inherent stability guarantee. To overcome this disadvantage, an a priori stability condition based on a terminal weighting matrix, i.e. enforcing that the value function is a Lyapunov function, a posteriori stability criteria based on constructing a PWQ Lyapunov function and an a posteriori stability criterion based on considering the value function as a candidate Lyapunov function, i.e. checking whether the value function is a Lyapunov function, have been proposed. These conditions and criteria differ in terms of conservatism and applicability for relaxed receding-horizon control and scheduling (RRHCS) and switched systems with non-stabilizable (non-stab.) subsystems as indicated in Figure 5.1. The a posteriori stability criterion based on a PWQ Lyapunov function considering the regionality given in Theorem 3.8 is least conservative as shown in a numerical study while the a posteriori stability criterion based on the value function as a candidate Lyapunov function given in Theorem 3.9 is most widely applicable. The choice of a condition or criterion therefore depends on the application.

The main advantage of periodic control and scheduling consists in an inherent stability guarantee. The main disadvantage consists in the lack of a complexity reduction during optimization. Ideas to overcome this disadvantage are proposed in Section 5.2.

The advantages and disadvantages of receding-horizon and periodic control and scheduling are obviously complementary. Therefore, for most applications an appropriate method is readily available. The methods are highly formalized, either analytically or as LMIs, and algorithms for an efficient implementation are provided, which is crucial in practice. The selection of a specific method ultimately depends on the application. A comparison of the methods in terms of performance is given in the case study presented in Chapter 10.

5.2 Future Work

The framework can be extended in various directions. Ideas for future work include

Infinite-Horizon Control and Scheduling

Infinite-horizon strategies are very attractive since closed-loop stability is commonly guaranteed inherently. For switched systems general conditions under which closed-loop stability is ensured under an infinite-horizon strategy still need to be established. The cornerstones are conditions for reachability and observability of switched systems and the convergence of the value iteration as pointed out in Section 1.4.3. Fundamental research is indispensable here. A good basis may be the results in [ZHA09]. Generalizing these results for switched systems with non-stabilizable subsystems could be a reasonable direction of future research. Furthermore, the results presented in [BCM06] for PWA systems and a linear cost function may be reconsidered for switched systems and a quadratic cost function. It must, however, be regarded that the optimization problem for determining an infinite-horizon strategy may often be computationally intractable, particularly for switched systems with a large number of subsystems M.

Robust Control and Scheduling

Robustness is crucial in control since uncertainties, disturbances and noise are always present. Robust control and scheduling is therefore an important direction of future research. Particularly approaches for robust RHC of linear systems [KBM96, SM98, BBM03, LCRM04, etc.] and PWA systems [KM02, Laz06, SKMJ08, etc.] may be a firm basis for extending receding-horizon control and scheduling and also periodic control and scheduling w.r.t. robustness.

A concept for robust periodic control and online scheduling (Robust $\mathrm{PCS}_{\mathrm{on}}$) of discrete-time switched linear systems with polytopic and norm-bounded uncertainty has been recently proposed in [AGL11]. The periodic control subproblem is addressed based on a periodic parameter-dependent Lyapunov function (LMI optimization problem), the periodic scheduling subproblem based on exhaustive search. Similar concepts may be sought for robust receding-horizon control and scheduling.

Suboptimality Analysis

Receding-horizon control and scheduling as well as periodic control and scheduling can be considered as an approximation of infinite-horizon control and scheduling. Quantifying the suboptimality induced by this approximation is certainly desirable as pointed out in Remark 3.2. To this end, suboptimality criteria for RHC of linear and nonlinear systems [NP97, PN00, GR08, Grü09] may be reconsidered for receding-horizon control and scheduling and periodic control and scheduling of switched systems. These criteria generally rely on controllability and observability conditions. Such conditions still need to be established for switched systems. Noteworthy, if a finite suboptimality bound can be found, then closed-loop stability is guaranteed. Thus, suboptimality is strongly related to stability which may be exploited for stability analysis.

Efficient Storage and Evaluation of the PWL State Feedback Control Law

The complexity for storing and evaluating the PWL state feedback control law resulting under RHCS, RRHCS, $\mathrm{PCS}_{\mathrm{on}}$, $\mathrm{RPCS}_{\mathrm{on}}$, $\mathrm{HPCS}_{\mathrm{on}}$ and OPP is characterized by $\mathcal{O}(n_r)$, where n_r is the number of regions. Particularly for switchted systems with fast dynamics, a high order n and a large number of subsystems M, this complexity may be critical.

Methods for reducing the storage and evaluation complexity have been intensely studied for explicit RHC of linear and PWA systems, see [AB09, Section 4] for a survey. E.g. in [TJB03], a method to reduce the complexity for evaluating a PWA state feedback control law from $\mathcal{O}(n_r)$ to $\mathcal{O}(\log n_r)$ has been proposed, indicating the potential of such methods. The proposed methods are tailored to PWA state feedback control laws defined over polyhedral regions. Unfortunately, a reformulation for PWL state feedback control laws defined over regions which are characterized by quadratic forms is not straightforward. Nevertheless, the methods may be a basis for further research.

The complexity for evaluating the PWL state feedback control law can also be reduced by exploiting the reachability. For each region, the regions which are reachable within one time instant can be determined utilizing the non-reachability criterion proposed in Theorem 3.11. The region membership test can then be limited to the regions which are reachable from the current region. In this way, the complexity may be reduced significantly. This approach is, however, infeasible if the switched system is subject to state disturbances. For this setup further research is necessary.

Stability Conditions for Receding-Horizon Control and Scheduling

Since closed-loop stability is not guaranteed inherently under RHCS and RRHCS, a priori stability conditions and a posteriori stability criteria are crucial. Various a priori conditions and a posteriori stability criteria have been proposed in Section 3.3 and the effectiveness has been shown by a numerical study. Nevertheless, some issues remain:

An a priori stability condition for RRHCS can not be established based on enforcing the relaxed value function to be a Lyapunov function as outlined at the end of Section 3.3.1. Several alternative approaches can be considered:

- Various a priori stability conditions have been proposed for suboptimal RHC of linear systems [BF03, JG03, etc.] and PWA systems [SKMJ08, LH09, etc.]. There, the suboptimality is commonly specified in an additive manner. Therefore, these a priori stability conditions are not directly applicable for RRHCS since here the suboptimality is specified in a multiplicative manner. Though, some ideas may be transferable.
- For enforcing the relaxed value function to be a Lyapunov function, an additional constraint can be added to Problem 3.1 as proposed in [SMR99, Section IV]. Similar concepts have been formulated in terms of control Lyapunov functions [PND98, JYH99, JH05, GMTT05]. The complexity for solving Problem 3.1 may, however, be significantly increased by adding an additional constraint. Specifically, relaxed dynamic programming will probably not lead to a DRE.

- Considerable research in recent years has focused on selecting the prediction horizon such that closed-loop stability is guaranteed inherently, see the discussion at the end of Section 3.3.2. Transferring these ideas to RRHCS seems very promising.

The a priori stability condition for RHCS proposed in Theorems 3.3 and 3.4 is limited to the switched systems with stabilizable subsystems as outlined in Remark 3.10. This limitation may be overcome by considering not only a suboptimal state feedback control law but also a suboptimal switching law at the final predicted time instant as discussed in Remark 3.10. Further pursuing this idea seems promising.

Furthermore, alternative a posteriori stability criteria may be studied. To this end, the concepts summarized in [BGLM05, LH09] can be revisited.

Parametrization of the Relaxed Value Function

For parametrizing the relaxed value function, quadratic forms have been considered in Section 2.4, cf. (2.38). Although this parametrization is manifest for a quadratic cost function, other parametrizations may be studied to increase the effectiveness of relaxed dynamic programming, e.g. based on positive polynomials [Wer07], as pointed out in Remark 2.13. Also other methods for approximate dynamic programming, see [BT96, Ber05a, Ber05b, Ran06, Pow07] for an overview, may be inspected.

Complexity Reduction for Periodic Control and Scheduling

The offline complexity of PCS_{off}, PCS_{on}, RPCS_{on}, HPCS_{on} and OPP still needs to be addressed, cf. Section 5.1. Several approaches for reducing the offline complexity have been proposed in the literature. In [RS04], it has been shown that the search space for determining the optimal p-periodic switching sequence has an algebraic structure and a heuristic exploiting this algebraic structure to reduce the offline complexity has been proposed. In [LHB09], stochastic optimization algorithms such as genetic algorithms and particle swarm optimization have been studied to reduce the offline complexity. The approaches in [RS04] and [LHB09] both guarantee closed-loop stability. Optimality is not ensured; however, good performance could be attested in numerical studies. In [BÇH09], a two-step procedure is proposed: First, the optimal p-periodic switching sequence is determined based on mixed-integer programming w.r.t. a finite-horizon $\mathcal{H}_2$ norm for periodic systems. Then, the optimal p-periodic state feedback control law associated to the optimal p-periodic switching sequence is computed based on periodic control theory. Closed-loop stability is guaranteed. Optimality is, however, not ensured in the strict sense due using a finite-horizon $\mathcal{H}_2$ norm to determine the optimal p-periodic switching sequence. Note that a similar two-step procedure can be formulated utilizing relaxed dynamic programming with periodicity constraints and a finite time horizon instead of mixed integer programming. Despite these important results, further research on a complexity reduction with guaranteed stability and performance is required.

One approach may consist in relaxed dynamic programming with periodicity constraints and an infinite time horizon. The convergence of the relaxed value iteration is again a crucial issue here.

Part II

Optimal Control of Networked Embedded Control Systems

6 Introduction

6.1 Networked Embedded Control Systems

Networked embedded control systems (NECSs) are control systems where controllers, sensors and actuators are connected via a communication network and where controllers are implemented on processors which are embedded into the application. The structure of an NECS is shown in Figure 6.1. NECSs occur in various applications, ranging from automobiles and aircrafts [JTN05] over manufacturing and process control [MT07] to power systems [LGOH09]. For example, modern cars contain up to 70 electronic control units and several bus systems such as CAN, LIN and FlexRay [NHL05]. Most of these are dedicated to control such as engine control, traction control, electronic stability control, anti-lock braking systems and climate control.

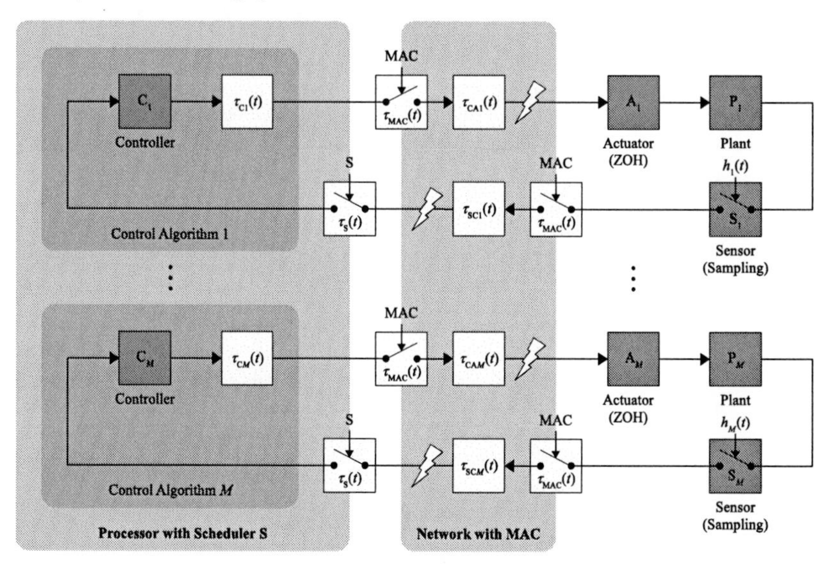

Figure 6.1: Structure of an NECS

NECSs provide several advantages over conventional control systems. On the one hand, costs can be reduced. Specifically wiring and maintenance costs can be considerably decreased while reusability and reconfigurability can be increased. Furthermore, fewer and better utilized processors are required. On the other hand, the functionality can be extended. For example, the control of mobile systems and spatially distributed systems is enabled over wireless networks and wide area networks. Furthermore, all information available over the network can be utilized for control. This additional information allows for completely new control concepts such as sensor-actuator networks.

Computation, communication and control have traditionally been addressed separately. The control community has focused on sampled-data control theory based on constant sampling periods and negligible or constant computation and transmission times (latencies) [ÅW90, FPW90], assuming that these are guaranteed by the computing and communication system. The computing and communication community has focused on real-time scheduling algorithms and medium access control (MAC) protocols based on periodic tasks and messages with fixed period, worst-case computation and transmission time and hard deadline [But05], assuming that these are needed by the control algorithm. This separation of concerns has proved very effective but also has drawbacks. First, meeting deadlines is not exactly equivalent to ensuring constant sampling periods and latencies. As a consequence, sampling periods and latencies are subject to variations (jitter) under standard real-time scheduling algorithms such as rate-monotonic (RM) and earliest-deadline first (EDF) scheduling and medium access control protocols such as carrier sense multiple access with collision avoidance (CSMA/CA, e.g. used in CAN). For example, sampling may be postponed and latencies may be extended under RM and EDF scheduling due to preemptions while deadlines are still met as illustrated in [GIL07]. These variations can influence performance or even stability [CHL+03, CvdWHN09]. Second, real-time scheduling algorithms and medium access control protocols partly rely on conservative assumptions such as worst-case computation and transmission times. As a consequence, computation and communication resources are not utilized efficiently. Furthermore, expensive specialized hardware and software is partly required. Third, wireless networks are increasingly employed where constant latencies are very difficult to guarantee.

There is a strong trend towards more flexible computation, communication and control concepts. These concepts aim at utilizing computation and communication resources more efficiently, employing inexpensive commercial-of-the-shelf (COTS) hardware and software and realizing novel functionalities. Migrating from hard real-time systems and specialized hardware and software to soft real-time systems and COTS hardware and software, however, significantly increases and also induces computation and communication imperfections, cf. Figure 6.1, specifically:

- *Computation Times* $\tau_{\mathrm{C}i}(t)$ *and Transmission Times* $\tau_{\mathrm{CA}i}(t)/\tau_{\mathrm{SC}i}(t)$

  Computation and transmission times result from the actual computation of the control law and the actual physical transmission of the data packet. They depend on the control algorithm, packet size, data rate, wire length etc. The computation times may be uncertain and time-varying due to pipelining, caching, preemptions etc., the transmission times due to varying packet size, routing etc. From a control perspective, computation and transmission times can be considered as time delays.

- *Packet Loss*

  Packet loss (indicated by flashes in Figure 6.1) results from transmission errors over the physical medium, which is particularly pronounced in wireless networks, and congestion. Moreover, routing may cause packet reordering, resulting in outdated and therefore lost packets. From a control perspective, packet loss can lead e.g. to time-varying sampling periods $h_i(t)$.

- *Access Times* $\tau_{\mathrm{S}}(t)$ *and* $\tau_{\mathrm{MAC}}(t)$

  Access times result from contention for shared computation and communication resources. They may be uncertain and time-varying depending on the scheduling algorithm and medium access control protocol. From a control perspective, access times can be considered as time delays and may also lead to uncertain time-varying sampling periods $h_i(t)$, e.g. if sampling is triggered by the control algorithm.

- *Quantization Effects*

  Quantization effects result from analog-to-digital/digital-to-analog conversion with limited resolution. These effects are always present in sampled-data control and have been widely studied, see e.g. [ÅW90, Section 15.6], [FPW90, Chapter 7], but become much more pronounced under data rate constraints necessitating a low quantization or even a lossy data compression.

Traditional sampled-data control theory which relies on constant sampling periods and negligible or constant latencies is no longer applicable under these computation and communications imperfections. Instead, novel methods for modeling, stability analysis and control design are required to ensure stability and performance. The integration of computation, communication and control is indeed considered as a key research direction of the future [MÅB+03].

Integrating computation, communication and control to NECSs has received considerable attention in recent years. This is witnessed by several surveys [TC03, Yan06, HNX07, BA07, Zam08], special issues [Bus01, AB04, AB07] and monographs [HL05, BHJ10]. Various concepts have been proposed which can be categorized into *implementation-aware control* and *control and scheduling codesign.*

6.2 Implementation-Aware Control

Implementation-aware control is focused on robustness with respect to uncertain time-varying computation, transmission and access times, packet loss and quantization effects.

This approach is mandatory if the computing and communication platform, particularly the scheduling algorithm and medium access control protocol, can not be influenced, which is usually the case in large heterogeneous NECSs based on COTS hardware and software. Implementation-aware control has been studied from various perspectives, including continuous-time, discrete-time and sampled-data control theory, deterministic and stochastic models as well as stability analysis and control design. Excellent overviews can be found in the surveys, special issues and monographs indicated above.

6.3 Control and Scheduling Codesign

Control and scheduling codesign is focused on an integrated design of the controller, scheduling algorithm and medium access control protocol. This approach is viable if the computing and communication platform can be influenced, which is commonly the case in small homogeneous NECSs. There are, however, attempts to apply control and scheduling codesign also for large heterogeneous NECSs based on COTS hardware and software by augmenting or superseding existing scheduling algorithms and medium access control protocols. Control and scheduling codesign provides substantial advantages over implementation-aware control: First, control and scheduling codesign exploits all degrees of freedom (computation, communication and control), while implementation-aware control only utilizes one degree of freedom (control). Second, control and scheduling codesign is proactive, i.e. computation and communication imperfections can be avoided or influenced, while implementation-aware control is reactive, i.e. computation and communication imperfections can only be tolerated or compensated. These advantages make control and scheduling codesign very promising for achieving both an efficient resource utilization and a high control performance.

The control and scheduling codesign problem can, with reference to the NECS structure given in Figure 6.1, be stated as

Problem 6.1 *Given*

- *a set of plants* $\mathbb{P} = \{\mathrm{P}_i, i \in \{1, \ldots, M\}\}$ *to be controlled,*
- *a processor with limited computation capacity and*
- *a network with limited communication capacity*

find

- *a set of control tasks* $\mathbb{T} = \{\mathrm{T}_i, i \in \{1, \ldots, M\}\}$ *including controllers* C_i *and*
- *a scheduler* S *and medium access control* MAC

such that the overall control performance is optimized.

Note that Problem 6.1 is partly reduced in the literature to finding only a scheduler S and medium access control MAC for given controllers C_i and to ensuring only stability.

Various concepts for control and scheduling codesign have been proposed in recent years, see [ÅC05, CA06, XS06, LVM07, LMVF08, XS08, SSA10] for an overview. Control and scheduling codesign has though been studied much less frequently than implementation-aware control. Concepts for control and scheduling codesign can be classified by

- *Offline Scheduling vs. Online Scheduling*

  Scheduling can be done offline (before runtime, OFF) or online (at runtime, ON). Taking scheduling decisions online requires computations and transmissions which induce scheduling overhead but also enables adaptivity under reference changes, disturbances, workload changes etc. From this perspective, an online scheduling with small scheduling overhead is desirable.

- *Centralized Scheduling vs. Decentralized Scheduling*

  Scheduling can be performed centralized (C) or decentralized (D). Centralized scheduling utilizes a central entity for taking scheduling decisions. A breakdown of this entity may cause a breakdown of the complete NECS. Centralized online scheduling furthermore usually requires global information for taking scheduling decisions. The transmission of this information to the central entity may induce significant scheduling overhead. Contention handling is not required or simple for centralized scheduling. Decentralized scheduling usually requires only local information for taking scheduling decisions. Contention handling is, however, rather complex, particularly from a control and scheduling codesign perspective.

- *Time-Triggered Scheduling vs. Event-Triggered Scheduling*

  Scheduling can be performed time-triggered (TT) or event-triggered (ET). Under time-triggered scheduling, scheduling decisions are taken at fixed time instants or variable time instants, e.g. when a computation or transmission has finished, while under event-triggered scheduling, scheduling decisions are taken at events, e.g. when the state exceeds a certain threshold.

- *Scheduling Criterion*

  Scheduling can be performed based on various criteria. If control and scheduling codesign is focused on stabilization, then scheduling is commonly based on instantaneous (inst.) quantities, such as the current control error, the current state, the $\mathcal{H}_\infty$ norm or the $\mathcal{L}_p$ norm. If control and scheduling codesign is focused on optimization, then scheduling is commonly based in cumulative (cum.) quantities, such as quadratic cost functions (LQ) or the $\mathcal{H}_2$ norm. Partly, approximate models and methods are considered in the literature.

- *Implementation*

  Scheduling can be implemented based on standard scheduling algorithms and medium access control protocols by adapting parameters like periods and priorities or on sequences.

Control and scheduling codesign can furthermore be classified by the usage of continuous-time, discrete-time or sampled-data control theory, the supported control strategies, the consideration of noise, the consideration of sampling and latency jitter, contention handling and whether the controllers are given or codesigned.

Various papers on control and scheduling codesign are listed chronologically and classified in Table 6.1 using the abbreviations defined above. This list is far from complete but reveals trends. There is a clear trend from offline scheduling to online scheduling, from time-triggered scheduling to event-triggered scheduling and from given controllers to controller codesign. On the one hand, centralized time-triggered online scheduling has been significantly advanced over the years. Particularly, the codesign of controllers and scheduling criteria based on optimization has been increasingly examined. On the other hand, decentralized event-triggered online scheduling has been intensely studied in recent years. The codesign of controllers, scheduling criteria based on optimization and implementation issues, however, still require research. Notably, scheduling criteria based on optimization have been rarely considered for networked control systems. Moreover, either embedded control systems (ECSs) or networked control systems (NCSs) have been studied while NECSs have not been addressed.

Paper	Scope	OFF/ON	C/D	TT/ET	Scheduling Criterion	Implementation	Controller
[SLSS96]	ECS	OFF	C	TT	cum. (LQ, approx.)	RM/EDF (periods)	given
[RS00, RS04, LHB09]	E/NCS	OFF	C	TT	cum. (LQ)	sequence	codesigned
[WY01]	NCS	ON	D	TT	inst. (control error)	CSMA/CA (priorities)	given
[Hri01, LZFA05]	NCS	OFF	C	TT	inst. (stability/$\mathcal{H}_\infty$)	sequence	given
[HC05]	ECS	ON	C	TT	cum. (LQ, approx.)	RM/EDF (periods)	given
[CA06]	ECS	ON	C	TT	cum. (LQ)	sequence	codesigned
[BÇH06]	NCS	ON	C	TT	cum. (LQ)	sequence	codesigned
[Tab07, WL09]	ECS	ON	D	ET	inst. (state/$\mathcal{L}_2$)	not addressed	given
[CH08, RJ09]	NCS	ON	D	ET	inst. (state)	basic concepts	given
[BÇH09]	ECS	ON/OFF	C	TT	cum. ($\mathcal{H}_2$/LQ)	sequence	codesigned
This thesis	NECS	ON/OFF	C	TT	cum. (LQ)	sequence	codesigned

Table 6.1: Classification of papers on control and scheduling codesign

The concept presented in this thesis is based on centralized time-triggered offline and online scheduling in the tradition of [RS00, RS04, CA06, BÇH06, BÇH09, LHB09].

Contributions

A general framework for control and scheduling codesign is presented in Chapters 7 to 10.

Modeling of NECSs is addressed in Chapter 7. An NECS architecture consisting of M plants controlled over a single network with medium access control by a single processor with scheduler is proposed in Section 7.1. For this NECS architecture a computation and communication model is devised in Section 7.2. Computation and communication are abstracted to tasks, scheduling and medium access control to a task sequence. This abstraction allows to integrate computation, communication and control. Moreover, an idle time is considered to accommodate further periodic, sporadic or aperiodic tasks. An NECS model corresponding to this computation and communication model is developed in Section 7.3. Each plant together with actuator and sensor is described by a continuous-time linear time-invariant state equation which contains the computation and transmission times as an input delay. The continuous-time state equation is discretized utilizing a special zero-order hold (ZOH) behavior: If the task associated to the plant under discretization is executed, then a new control vector is computed, transmitted and updated within the discretization interval; otherwise the preceding control vector is held over the discretization interval. After discretization each plant is described by an augmented discrete-time linear time-varying state equation. The discrete-time state equations of the individual plants are combined into a block-diagonal discrete-time state equation representing the overall NECS. For evaluating the performance of the NECS, each plant is assigned a continuous-time quadratic cost function which is discretized for the special ZOH behavior. In this way time-varying sampling periods, computation times and transmission times induced by scheduling and medium access control are reflected in the discretized cost function, rendering different task sequences comparable. The discretized cost functions for the individual plants are combined into a block-diagonal discrete-time cost function for the overall NECS. Related concepts have been proposed in [RS00, RS04, CA06, BÇH06, BÇH09, LHB09]. The contribution in Chapter 7 is twofold: First, the proposed concept integrates computation, communication and control, while previous concepts focus either on computation and control [RS00, CA06, BÇH09] or on communication and control [RS04, BÇH06, LHB09]. Second, the proposed concept allows to handle computation and communication times in a very flexible way. Arbitrary and different computation and transmission times can be considered for each task. In this way the timing characteristics of different control algorithms and for different packet sizes can be represented exactly. Previous concepts on the contrary rely on granularity of the computation and transmission times, with [CA06] as a notable exception.

Control and scheduling codesign of NECSs is addressed in Chapter 8. The control and scheduling codesign problem is formalized in Section 8.1 (Problem 8.1) based on the NECS model and cost function derived in Chapter 7. It is shown that the NECS model corresponds to a discrete-time switched linear system (1.3) with the switching index as interface to scheduling and medium access control and the control vector as interface to control. Control and scheduling codesign of NECSs (Problem 8.1) is then a special case of optimal control and scheduling of switched systems (Problem 1.9). The methods presented in Chapters 2 to 4 are therefore applicable for control and scheduling codesign. The methods support offline scheduling as well as online scheduling and allow balancing between complexity, both offline and online, and performance in a very systematic way. Furthermore, stability and performance are rigorously addressed. These issues have been considered only partly in previous work [RS00, RS04, CA06, BÇH06, BÇH09, LHB09], see also the discussion at the end of Section 5.2. The structural properties of the NECS model derived in Chapter 7 are analyzed in Section 8.2. It is revealed in Proposition 8.2 that the subsystems of the NECS model are not stabilizable which limits the applicability of the methods presented in Chapter 3 (cf. Remarks 3.10 and 3.17).

N'-step receding-horizon control and scheduling (N'-step RHCS) of NECSs is proposed in Chapter 9 to overcome this limitation. The idea consists in applying not the first element but the first N' elements of the control and task sequence to the NECS and thus in reoptimizing not at each time instant but each N' time instants. The dynamic programming solution, relaxed dynamic programming solution and explicit solution presented in Sections 2.2, 2.3, 2.4 and 3.2 are reconsidered for N'-step RHCS in Sections 9.2 and 9.4. A posteriori stability criteria and a non-reachability criterion are formulated in Sections 9.5 and 9.6 based on lifting over N' time instants. Furthermore, suboptimality analysis is addressed in Section 9.3. Similar strategies have not been considered before.

A case study is given in Chapter 10 where modeling and control and scheduling codesign are illustrated and evaluated for networked control of three inverted pendulums.

7 Modeling

7.1 NECS Architecture

Consider the NECS shown in Figure 7.1.

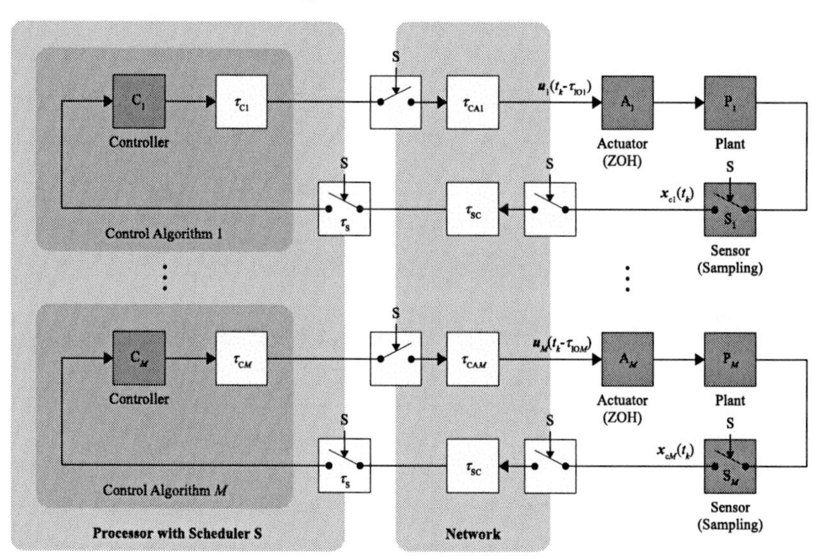

Figure 7.1: NECS Architecture

The NECS consists of a set of plants

$$\mathbb{P} = \{\mathrm{P}_i, i \in \{1, \dots, M\}\} \tag{7.1}$$

controlled by a set of control tasks

$$\mathbb{T} = \{\mathrm{T}_i, i \in \{1, \dots, M\}\} \tag{7.2}$$

where plant P_i is controlled by control tasks T_i. The actuators A_i contain a zero-order hold (ZOH) device and the sensors S_i a sampling device.

The tasks are implemented on a single processor and moreover the control signals $\boldsymbol{u}_i(t_k)$ and sensor signals $\boldsymbol{x}_{ci}(t_k)$ are transmitted over a shared network. Therefore, a scheduling of the processor and a medium access control of the network are required. Scheduling and medium access control are realized jointly by a centralized scheduler S which is implemented on the processor.

It is assumed that no packet loss occurs and that the quantization effects are negligible.

In the following section a computation and communication model for this NECS is introduced.

7.2 Computation and Communication Model

The computation and communication are controlled by the scheduler S which generates a task sequence or schedule. The task sequence indicates the execution order of the tasks and is formally defined by

Definition 7.1 *A task sequence* $j : \mathbb{N}_0 \to \mathbb{M} = \{1, \ldots, M\}$ *is defined by*

$$j(k) = i \text{ if task } \mathrm{T}_i \text{ is executed in time interval } t_k \leq t < t_{k+1} \text{ with } k \in \mathbb{N}_0. \tag{7.3}$$

The computation and communication proceeds in five steps:

1. Sample and transmit the state vectors $\boldsymbol{x}_{ci}(t_k)$ of all plants $\mathrm{P}_i \in \mathbb{P}$
   (Sampling is triggered by the scheduler S for all plants $\mathrm{P}_i \in \mathbb{P}$ simultaneously.)
   (Transmission time $\tau_{\mathrm{SC}} = \sum_{i=1}^{M} \tau_{\mathrm{SC}i} = \text{const.}$)
2. Determine the plant $\mathrm{P}_{j(k)} \in \mathbb{P}$ for which the control vector shall be computed and transmitted based on the state vectors $\boldsymbol{x}_{ci}(t_k)$ of all plants $\mathrm{P}_i \in \mathbb{P}$
   (Computation time $\tau_{\mathrm{S}} = \text{const.}$)
3. Compute the control vector $\boldsymbol{u}_{j(k)}(t_k)$ of plant $\mathrm{P}_{j(k)} \in \mathbb{P}$
   (Computation time $\tau_{\mathrm{C}j(k)} = \text{const.}$)
4. Transmit and update the control vector $\boldsymbol{u}_{j(k)}(t_k)$ of plant $\mathrm{P}_{j(k)} \in \mathbb{P}$
   (Transmission time $\tau_{\mathrm{CA}j(k)} = \text{const.}$)
5. Idle
   (Optional, e.g. for triggering sampling over the network (see step 1) or for performing other periodic, sporadic and aperiodic computations and transmissions)
   (Idle time $\tau_{\mathrm{I}j(k)} = \text{const.}$)

Steps 1 and 2 are related to the scheduler S whereas steps 3, 4 and 5 are related to the control task $\mathrm{T}_{j(k)}$.

The computation and communication model is illustrated in Figure 7.2 where sampling is indicated by a circle and control update by a square.

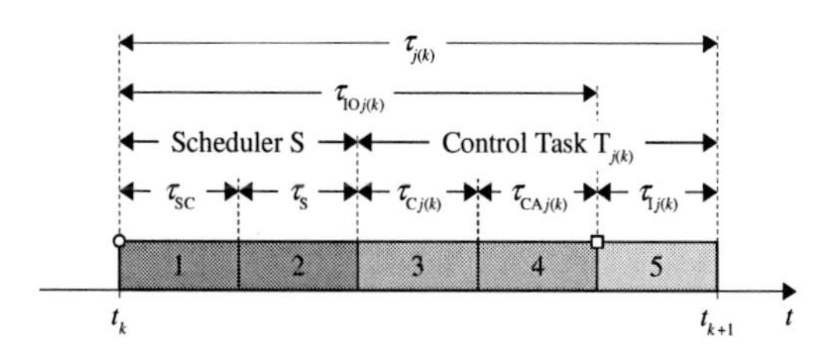

Figure 7.2: Computation and communication model

Remark 7.1. Scheduling according to this computation and communication model can be classified as online, centralized and time-triggered, cf. Section 6.3.

Remark 7.2. Step 1 may induce considerable scheduling overhead but is necessary since the state vectors $\boldsymbol{x}_{\mathrm{c}i}(t_k)$ of all plants $\mathrm{P}_i \in \mathbb{P}$ are required for online scheduling. Strategies for reducing the scheduling overhead can be based on observers and predictors to reconstruct missing state information, on hybrid scheduling, i.e. offline scheduling of the sensor-controller link and online scheduling of the controller-actuator link, and on distributed optimization, see Section 11.2 for a detailed discussion. The scheduling overhead does not arise in ECS ($\tau_{\mathrm{SC}i} = \tau_{\mathrm{CA}i} = 0$) or in NECSs where sensors and controllers are not connected over a network ($\tau_{\mathrm{SC}i} = 0$).

The input-output delay, i.e. the time between sampling and control update, is given by

$$\tau_{\mathrm{IO}j(k)} = \tau_{\mathrm{SC}} + \tau_{\mathrm{S}} + \tau_{\mathrm{C}j(k)} + \tau_{\mathrm{CA}j(k)}, \tag{7.4}$$

the overall delay by

$$\tau_{j(k)} = \tau_{\mathrm{IO}j(k)} + \tau_{\mathrm{I}j(k)}. \tag{7.5}$$

Both obviously depend on the task sequence $j(k)$ and are therefore time-varying.

In the following section an NECS model corresponding to the computation and communication model is derived.

7.3 NECS Model

Each plant $\mathrm{P}_i \in \mathbb{P}$ including actuator A_i and sensor S_i is described by a continuous-time linear time-invariant state equation

$$\begin{aligned} \dot{\boldsymbol{x}}_{\mathrm{c}i}(t) &= \boldsymbol{A}_{\mathrm{c}i}\boldsymbol{x}_{\mathrm{c}i}(t) + \boldsymbol{B}_{\mathrm{c}i}\boldsymbol{u}_i(t - \tau_{\mathrm{IO}i}) \\ \boldsymbol{x}_{\mathrm{c}i}(0) &= \boldsymbol{x}_{\mathrm{c}i0} \end{aligned} \tag{7.6}$$

where $\boldsymbol{A}_{\mathrm{c}i} \in \mathbb{R}^{\mathrm{n}_i \times \mathrm{n}_i}$ is the system matrix, $\boldsymbol{B}_{\mathrm{c}i} \in \mathbb{R}^{\mathrm{n}_i \times \mathrm{m}_i}$ is the input matrix, $\boldsymbol{x}_{\mathrm{c}i} \in \mathbb{R}^{\mathrm{n}_i}$ is the state vector and $\boldsymbol{u}_i \in \mathbb{R}^{\mathrm{m}_i}$ is the control vector with the input delay $\tau_{\mathrm{IO}i}$ corresponding to the input-output delay.

The continuous-time state equation (7.6) is discretized over the discretization interval $t_k \leq t < t_{k+1}$ using ZOH. The discretization interval is characterized by the overall delay $\tau_{j(k)}$ of the executed task $\mathrm{T}_{j(k)}$, i.e.

$$t_{k+1} - t_k = \tau_{j(k)}. \tag{7.7}$$

The sampling period $h_i(k_i)$ of a task T_i on the contrary is determined by the sum of the overall delays $\tau_{j(k)}$ of the tasks $\mathrm{T}_{j(k)}$ being executed within this sampling period. Note that the sampling periods $h_i(k_i)$ have to be indexed separately for each task T_i by an index $k_i \in \mathbb{N}_0$.

Example 7.1 Consider an NECS with two plants P_1 and P_2 controlled by two control tasks T_1 and T_2. Furthermore, inspect the partial schedule $j(k) = (\ldots, 1, 2, 2, 1, 2, 1, \ldots)$. The sampling periods $h_i(k_i)$ of the individual tasks T_i are then given by

$$\begin{aligned} h_1(k_1) &= \tau_1 + \tau_2 + \tau_2 \\ h_1(k_1+1) &= \tau_1 + \tau_2 \\ h_2(k_2) &= \tau_2 \\ h_2(k_2+1) &= \tau_2 + \tau_1. \end{aligned}$$

An illustration of the timing of the NECS is given in Figure 7.3.

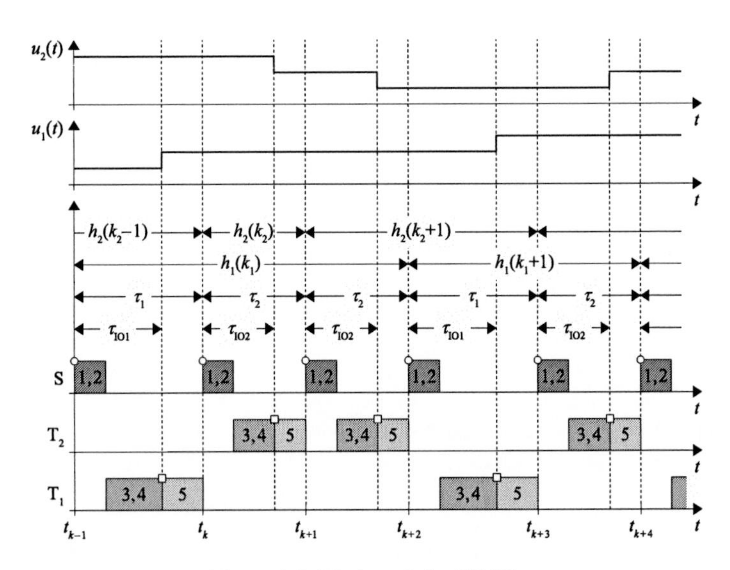

Figure 7.3: Timing of the NECS

□

The sampling periods $h_i(k_i)$ of the individual tasks T_i obviously depend on the task sequence $j(k)$ and are consequently time-varying.

For the discretization, the task T_i considered for discretization and the task $\mathrm{T}_{j(k)}$ executed within the discretization interval $t_k \leq t < t_{k+1}$ must be distinguished. If the considered task is the executed task ($i = j(k)$), then the control vector is updated within the discretization interval, i.e.

$$\boldsymbol{u}_i(t) = \begin{cases} \boldsymbol{u}_i(t_{k-1}) & \text{for} \quad t_k \leq t < t_k + \tau_{\mathrm{IO}i} \\ \boldsymbol{u}_i(t_k) & \text{for} \quad t_k + \tau_{\mathrm{IO}i} \leq t < t_{k+1}. \end{cases} \tag{7.8}$$

If the considered task is not the executed task ($i \neq j(k)$), then the control vector is not updated at all, i.e.

$$\boldsymbol{u}_i(t) = \boldsymbol{u}_i(t_{k-1}) \quad \text{for} \quad t_k \leq t < t_{k+1}. \tag{7.9}$$

This specific ZOH behavior is illustrated in Figure 7.3.

In the following the distinction between the considered task and the executed task is represented by the logical variable

$$\delta_{ij(k)} = \begin{cases} 1 & \text{if} \quad i = j(k) \\ 0 & \text{if} \quad i \neq j(k). \end{cases} \tag{7.10}$$

This allows to define a generalized delay

$$\tau_{ij(k)} = \delta_{ij(k)}\tau_{\mathrm{IO}i} + (1 - \delta_{ij(k)})\tau_{j(k)}. \tag{7.11}$$

Equations (7.8) and (7.9) can then be combined to

$$\boldsymbol{u}_i(t) = \begin{cases} \boldsymbol{u}_i(t_{k-1}) & \text{for} \quad t_k \leq t < t_k + \tau_{ij(k)} \\ \boldsymbol{u}_i(t_k) & \text{for} \quad t_k + \tau_{ij(k)} \leq t < t_{k+1}. \end{cases} \tag{7.12}$$

For this specific ZOH behavior the solution of the continuous-time state equation (7.6) at a time $t \geq t_k$ for a given state vector $\boldsymbol{x}(t_k)$ is given by

$$\boldsymbol{x}_{\mathrm{c}i}(t) = \mathrm{e}^{\boldsymbol{A}_{\mathrm{c}i}(t-t_k)}\boldsymbol{x}_{\mathrm{c}i}(t_k) + \int_{t_k}^{t} \mathrm{e}^{\boldsymbol{A}_{\mathrm{c}i}(t-s)}\boldsymbol{B}_{\mathrm{c}i}\boldsymbol{u}_i(s - \tau_{ij(k)})ds. \tag{7.13}$$

The solution over the discretization interval $t_k \leq t < t_{k+1}$ results by setting $t = t_{k+1}$.

The control vector $\boldsymbol{u}_i(t)$ is piecewise constant when applying ZOH. Hence, the delayed control vector $\boldsymbol{u}_i(s - \tau_{ij(k)})$ is also piecewise constant. However, the delayed control vector changes for $\delta_{ij(k)} = 1$ not at but between the discretization instants t_k and t_{k+1}, see (7.12) and Figure 7.3. The integral in (7.13) can therefore be split, leading to

$$\begin{aligned} \boldsymbol{x}_{\mathrm{c}i}(t_{k+1}) &= \mathrm{e}^{\boldsymbol{A}_{\mathrm{c}i}(t_{k+1}-t_k)}\boldsymbol{x}_{\mathrm{c}i}(t_k) \\ &+ \int_{t_k}^{t_k+\tau_{ij(k)}} \mathrm{e}^{\boldsymbol{A}_{\mathrm{c}i}(t_{k+1}-s)}ds\boldsymbol{B}_{\mathrm{c}i}\boldsymbol{u}_i(t_{k-1}) + \int_{t_k+\tau_{ij(k)}}^{t_{k+1}} \mathrm{e}^{\boldsymbol{A}_{\mathrm{c}i}(t_{k+1}-s)}ds\boldsymbol{B}_{\mathrm{c}i}\boldsymbol{u}_i(t_k) \end{aligned} \tag{7.14}$$

or after substituting (7.7) to

$$\begin{aligned}\boldsymbol{x}_{\mathrm{c}i}(k+1) =& \boldsymbol{\Phi}_{ij(k)}\left(\tau_{j(k)}\right)\boldsymbol{x}_{\mathrm{c}i}(k) \\ & + \boldsymbol{\Gamma}_{1ij(k)}\left(\tau_{j(k)}, \tau_{ij(k)}\right)\boldsymbol{u}_i(k-1) + \boldsymbol{\Gamma}_{0ij(k)}\left(\tau_{j(k)}, \tau_{ij(k)}\right)\boldsymbol{u}_i(k) \\ \boldsymbol{x}_{\mathrm{c}i}(0) =& \boldsymbol{x}_{\mathrm{c}i0}\end{aligned} \tag{7.15}$$

with

$$\boldsymbol{\Phi}_{ij(k)}\left(\tau_{j(k)}\right) = \mathrm{e}^{\boldsymbol{A}_{\mathrm{c}i}\tau_{j(k)}} \tag{7.16a}$$

$$\boldsymbol{\Gamma}_{1ij(k)}\left(\tau_{j(k)}, \tau_{ij(k)}\right) = \mathrm{e}^{\boldsymbol{A}_{\mathrm{c}i}(\tau_{j(k)}-\tau_{ij(k)})}\int_0^{\tau_{ij(k)}} \mathrm{e}^{\boldsymbol{A}_{\mathrm{c}i}s}ds\boldsymbol{B}_{\mathrm{c}i} \tag{7.16b}$$

$$\boldsymbol{\Gamma}_{0ij(k)}\left(\tau_{j(k)}, \tau_{ij(k)}\right) = \int_0^{\tau_{j(k)}-\tau_{ij(k)}} \mathrm{e}^{\boldsymbol{A}_{\mathrm{c}i}s}ds\boldsymbol{B}_{\mathrm{c}i} \tag{7.16c}$$

where the discrete time $k \in \mathbb{N}_0$ represents the discretization instants t_k in an abstract way. The computation of the matrices $\boldsymbol{\Gamma}_{1ij(k)}\left(\tau_{j(k)}, \tau_{ij(k)}\right)$ and $\boldsymbol{\Gamma}_{0ij(k)}\left(\tau_{j(k)}, \tau_{ij(k)}\right)$ can be reformulated into the computation of a matrix exponential as shown in [Van78]. Methods for computing matrix exponentials numerically have in turn been proposed in [MV78].

Note that if $\tau_{\mathrm{I}i} = 0$ for all control tasks $\mathrm{T}_i \in \mathbb{T}$, then the overall delays become $\tau_i = \tau_{\mathrm{IO}i}$ and the matrices (7.16b) and (7.16c) reduce to

$$\boldsymbol{\Gamma}_{1ij(k)}\left(\tau_{j(k)}\right) = \int_0^{\tau_{j(k)}} \mathrm{e}^{\boldsymbol{A}_{\mathrm{c}i}s}ds\boldsymbol{B}_{\mathrm{c}i} \tag{7.17a}$$

$$\boldsymbol{\Gamma}_{0ij(k)} = \boldsymbol{0}. \tag{7.17b}$$

The discrete-time state equation related to the continuous-time state equation (7.6) results by augmenting the state vector $\boldsymbol{x}_{\mathrm{c}i}(k)$ with the preceding control vector $\boldsymbol{u}_i(k-1)$ and including the logical variable $\delta_{ij(k)}$ as

$$\begin{aligned}\boldsymbol{x}_i(k+1) &= \boldsymbol{A}_{ij(k)}\boldsymbol{x}_i(k) + \boldsymbol{B}_{ij(k)}\boldsymbol{u}_i(k) \\ \boldsymbol{x}_i(0) &= \begin{pmatrix}\boldsymbol{x}_{\mathrm{c}i0}^T & \boldsymbol{0}\end{pmatrix}^T\end{aligned} \tag{7.18}$$

where the state vector $\boldsymbol{x}_i(k) \in \mathbb{R}^{\mathrm{n}_i+\mathrm{m}_i}$, the system matrix $\boldsymbol{A}_{ij(k)} \in \mathbb{R}^{(\mathrm{n}_i+\mathrm{m}_i)\times(\mathrm{n}_i+\mathrm{m}_i)}$ and the input matrix $\boldsymbol{B}_{ij(k)} \in \mathbb{R}^{(\mathrm{n}_i+\mathrm{m}_i)\times\mathrm{m}_i}$ are given by

$$\boldsymbol{x}_i(k) = \begin{pmatrix}\boldsymbol{x}_{\mathrm{c}i}(k) \\ \boldsymbol{u}_i(k-1)\end{pmatrix} \tag{7.19a}$$

$$\boldsymbol{A}_{ij(k)} = \begin{pmatrix}\boldsymbol{\Phi}_{ij(k)}\left(\tau_{j(k)}\right) & \boldsymbol{\Gamma}_{1ij(k)}\left(\tau_{j(k)}, \tau_{ij(k)}\right) \\ \boldsymbol{0} & \left(1-\delta_{ij(k)}\right)\boldsymbol{I}\end{pmatrix} \tag{7.19b}$$

$$\boldsymbol{B}_{ij(k)} = \begin{pmatrix}\boldsymbol{\Gamma}_{0ij(k)}\left(\tau_{j(k)}, \tau_{ij(k)}\right) \\ \delta_{ij(k)}\boldsymbol{I}\end{pmatrix}. \tag{7.19c}$$

For $\delta_{ij(k)} = 1$, the subsequent state vector $\boldsymbol{x}_i(k+1)$ is affected by the preceding control vector $\boldsymbol{u}_i(k-1)$ via $\boldsymbol{\Gamma}_{1ij(k)}\left(\tau_{j(k)}, \tau_{\mathrm{IO}j(k)}\right)$ relating to $t_k \leq t < t_k + \tau_{\mathrm{IO}j(k)}$ and by the current control vector $\boldsymbol{u}_i(k)$ via $\boldsymbol{\Gamma}_{0ij(k)}\left(\tau_{j(k)}, \tau_{\mathrm{IO}j(k)}\right)$ relating to $t_k + \tau_{\mathrm{IO}j(k)} \leq t < t_{k+1}$. Furthermore, the current control vector $\boldsymbol{u}_i(k)$ is stored into the subsequent state vector $\boldsymbol{x}_i(k+1)$.

For $\delta_{ij(k)} = 0$, the subsequent state vector $\boldsymbol{x}_i(k+1)$ is affected by the preceding control vector $\boldsymbol{u}_i(k-1)$ via $\boldsymbol{\Gamma}_{1ij(k)}\left(\tau_{j(k)}, \tau_{j(k)}\right)$ relating to $t_k \leq t < t_k + \tau_{j(k)} = t_{k+1}$ but not by the current control vector $\boldsymbol{u}_i(k)$ since $\boldsymbol{\Gamma}_{0ij(k)}\left(\tau_{j(k)}, \tau_{j(k)}\right) = \mathbf{0}$. Furthermore, the preceding control vector $\boldsymbol{u}_i(k-1)$ stored in the current state vector $\boldsymbol{x}_i(k)$ is restored in subsequent state vector $\boldsymbol{x}_i(k+1)$.

Remark 7.3. The representation (7.18)/(7.19) generalizes a representation for time-delay systems proposed by [ÅW90, Section 3.2] to NECSs.

Remark 7.4. The control vector is held until a new control vector is available, cf. Figure 7.3. Alternatively, the control vector $\boldsymbol{u}_i(k)$ can be held only over $t_k + \tau_{\mathrm{IO}j(k)} \leq t \leq t_{k+2}$ and then set to zero until a new control vector is available. The system matrix (7.19b) is then changed to

$$\boldsymbol{A}_{ij(k)} = \begin{pmatrix} \boldsymbol{\Phi}_{ij(k)}\left(\tau_{j(k)}\right) & \boldsymbol{\Gamma}_{1ij(k)}\left(\tau_{j(k)}, \tau_{ij(k)}\right) \\ \mathbf{0} & \mathbf{0} \end{pmatrix} \tag{7.20}$$

while the input matrix (7.19c) is not changed. Selecting the ZOH behavior according to either (7.19b) or (7.20) very much depends on the NECS parameters, see [Hri07, Sch09] for a related discussion. For brevity only the ZOH behavior according to (7.19b) is considered in the following.

The NECS can be represented by a block-diagonal discrete-time state equation

$$\begin{aligned} \boldsymbol{x}(k+1) &= \boldsymbol{A}_{j(k)}\boldsymbol{x}(k) + \boldsymbol{B}_{j(k)}\boldsymbol{u}(k) \\ \boldsymbol{x}(0) &= \left(\boldsymbol{x}_1^T(0) \quad \cdots \quad \boldsymbol{x}_M^T(0)\right)^T \end{aligned} \tag{7.21}$$

where the state vector $\boldsymbol{x}(k) \in \mathbb{R}^{\mathrm{n}}$, the control vector $\boldsymbol{u}(k) \in \mathbb{R}^{\mathrm{m}}$, the system matrix $\boldsymbol{A}_{j(k)} \in \mathbb{R}^{\mathrm{n}\times\mathrm{n}}$ and the input matrix $\boldsymbol{B}_{j(k)} \in \mathbb{R}^{\mathrm{n}\times\mathrm{m}}$ with

$$\mathrm{n} = \sum_{i=1}^{M}(\mathrm{n}_i + \mathrm{m}_i), \quad \mathrm{m} = \sum_{i=1}^{M}\mathrm{m}_i \tag{7.22}$$

are given by

$$\boldsymbol{x}(k) = \begin{pmatrix} \boldsymbol{x}_1(k) \\ \vdots \\ \boldsymbol{x}_M(k) \end{pmatrix}, \quad \boldsymbol{u}(k) = \begin{pmatrix} \boldsymbol{u}_1(k) \\ \vdots \\ \boldsymbol{u}_M(k) \end{pmatrix} \tag{7.23a}$$

$$\boldsymbol{A}_{j(k)} = \begin{pmatrix} \boldsymbol{A}_{1j(k)} & \cdots & 0 \\ \vdots & \ddots & \vdots \\ 0 & \cdots & \boldsymbol{A}_{Mj(k)} \end{pmatrix}, \quad \boldsymbol{B}_{j(k)} = \begin{pmatrix} \boldsymbol{B}_{1j(k)} & \cdots & 0 \\ \vdots & \ddots & \vdots \\ 0 & \cdots & \boldsymbol{B}_{Mj(k)} \end{pmatrix}. \tag{7.23b}$$

This block-diagonal representation describes the dynamics of the overall NECS in the discretization interval $t_k \leq t < t_{k+1}$ with the control task $\mathrm{T}_{j(k)}$ being executed.

In the following section a cost function corresponding to the NECS model is introduced.

7.4 Cost Function

A cost function for control and scheduling codesign must respect both control and scheduling. Particularly, the time-varying sampling periods and time delays induced by scheduling must be considered in the cost function to render the costs for different task sequences comparable. For this purpose, a discrete-time time-invariant cost function is inappropriate. Instead, a continuous-time cost function must be imposed and discretized regarding the time-varying sampling periods and time delays which leads to a discrete-time time-varying cost function.

Consider the continuous-time quadratic cost function

$$J_{Ni} = \boldsymbol{x}_{\mathrm{c}i}^T(T)\boldsymbol{Q}_{0\mathrm{c}i}\boldsymbol{x}_{\mathrm{c}i}(T) + \int_0^T \boldsymbol{x}_{\mathrm{c}i}^T(t)\boldsymbol{Q}_{1\mathrm{c}i}\boldsymbol{x}_{\mathrm{c}i}(t) + \boldsymbol{u}_i^T(t-\tau_{\mathrm{IO}i})\boldsymbol{Q}_{2\mathrm{c}i}\boldsymbol{u}_i(t-\tau_{\mathrm{IO}i})dt \quad (7.24)$$

associated to plant P_i and control task T_i where $\boldsymbol{Q}_{0\mathrm{c}i} \in \mathbb{R}^{\mathrm{n}_i \times \mathrm{n}_i}$ and $\boldsymbol{Q}_{1\mathrm{c}i} \in \mathbb{R}^{\mathrm{n}_i \times \mathrm{n}_i}$ are symmetric and positive semidefinite weighting matrices, $\boldsymbol{Q}_{2\mathrm{c}i} \in \mathbb{R}^{\mathrm{m}_i \times \mathrm{m}_i}$ is a symmetric and positive definite weighting matrix and T is the time horizon.

The continuous-time quadratic cost function (7.24) can be discretized analogously to the continuous-time state equation (7.6). For the discretization, the considered task T_i and the executed task $\mathrm{T}_{j(k)}$ must again be distinguished. Therefor the concepts presented in Section 7.3 can be utilized. Details on the discretization are given in Appendix A.4.

The discrete-time quadratic cost function associated to the continuous-time quadratic cost function (7.24) results as

$$J_{Ni} = \boldsymbol{x}_i^T(N)\boldsymbol{Q}_{0i}\boldsymbol{x}_i(N) + \sum_{k=0}^{N-1}\left[\boldsymbol{x}_i^T(k)\boldsymbol{Q}_{1ij(k)}\boldsymbol{x}_i(k) + 2\boldsymbol{x}_i^T(k)\boldsymbol{Q}_{12ij(k)}\boldsymbol{u}_i(k) + \boldsymbol{u}_i^T(k)\boldsymbol{Q}_{2ij(k)}\boldsymbol{u}_i(k)\right] \quad (7.25)$$

where the discretized weighting matrices obey

$$\boldsymbol{Q}_{0i} = \begin{pmatrix} \boldsymbol{Q}_{0\mathrm{c}i} & \boldsymbol{0} \\ \boldsymbol{0} & \boldsymbol{0} \end{pmatrix} \quad (7.26a)$$

$$\boldsymbol{Q}_{1ij(k)} = \begin{pmatrix} \int_0^{\tau_{j(k)}} \boldsymbol{\Phi}_{ij(k)}^T(t)\boldsymbol{Q}_{1\mathrm{c}i}\boldsymbol{\Phi}_{ij(k)}(t)dt & \int_0^{\tau_{ij(k)}} \boldsymbol{\Phi}_{ij(k)}^T(t)\boldsymbol{Q}_{1\mathrm{c}i}\boldsymbol{\Gamma}_{ij(k)}(0,t)dt + \int_{\tau_{ij(k)}}^{\tau_{j(k)}} \boldsymbol{\Phi}_{ij(k)}^T(t)\boldsymbol{Q}_{1\mathrm{c}i}\boldsymbol{\Gamma}_{ij(k)}(0,\tau_{ij(k)})dt \\ * & \int_0^{\tau_{ij(k)}} \boldsymbol{\Gamma}_{ij(k)}^T(0,t)\boldsymbol{Q}_{1\mathrm{c}i}\boldsymbol{\Gamma}_{ij(k)}(0,t) + \boldsymbol{Q}_{2\mathrm{c}i}dt + \int_{\tau_{ij(k)}}^{\tau_{j(k)}} \boldsymbol{\Gamma}_{ij(k)}^T(0,\tau_{ij(k)})\boldsymbol{Q}_{1\mathrm{c}i}\boldsymbol{\Gamma}_{ij(k)}(0,\tau_{ij(k)})dt \end{pmatrix} \quad (7.26b)$$

$$\boldsymbol{Q}_{12ij(k)} = \begin{pmatrix} \int_{\tau_{ij(k)}}^{\tau_{j(k)}} \boldsymbol{\Phi}_{ij(k)}^T(t)\boldsymbol{Q}_{1\mathrm{c}i}\boldsymbol{\Gamma}_{ij(k)}(\tau_{ij(k)},t)dt \\ \int_{\tau_{ij(k)}}^{\tau_{j(k)}} \boldsymbol{\Gamma}_{ij(k)}^T(0,\tau_{ij(k)})\boldsymbol{Q}_{1\mathrm{c}i}\boldsymbol{\Gamma}_{ij(k)}(\tau_{ij(k)},t)dt \end{pmatrix} \quad (7.26c)$$

$$\boldsymbol{Q}_{2ij(k)} = \int_{\tau_{ij(k)}}^{\tau_{j(k)}} (\boldsymbol{\Gamma}_{ij(k)}^T(\tau_{ij(k)},t)\boldsymbol{Q}_{1\mathrm{c}i}\boldsymbol{\Gamma}_{ij(k)}(\tau_{ij(k)},t) + \boldsymbol{Q}_{2\mathrm{c}i}dt \quad (7.26d)$$

with

$$\boldsymbol{\Phi}_{ij(k)}(t) = \mathrm{e}^{\boldsymbol{A}_{\mathrm{c}i}t}, \quad \boldsymbol{\Gamma}_{ij(k)}(\underline{t},\bar{t}) = \int_{\underline{t}}^{\bar{t}} \mathrm{e}^{\boldsymbol{A}_{\mathrm{c}i}(t-s)} ds \boldsymbol{B}_{\mathrm{c}i}. \tag{7.27}$$

The weighting matrices $\boldsymbol{Q}_{0i} \in \mathbb{R}^{(\mathrm{n}_i+\mathrm{m}_i)\times(\mathrm{n}_i+\mathrm{m}_i)}$ and $\boldsymbol{Q}_{1ij(k)} \in \mathbb{R}^{(\mathrm{n}_i+\mathrm{m}_i)\times(\mathrm{n}_i+\mathrm{m}_i)}$ are always symmetric and positive semidefinite whereas the weighting matrix $\boldsymbol{Q}_{2ij(k)} \in \mathbb{R}^{\mathrm{m}_i\times\mathrm{m}_i}$ is symmetric and positive definite if the considered task is the executed task, i.e. $\delta_{ij(k)} = 1$, and zero if the considered task is not the executed task, i.e. $\delta_{ij(k)} = 0$. For $\delta_{ij(k)} = 0$ the control vector is not updated, i.e. no new control vector $\boldsymbol{u}_i(k)$ is computed, but the preceding control vector $\boldsymbol{u}_i(k-1)$ is held. The preceding control vector $\boldsymbol{u}_i(k-1)$ is in turn accounted for in the weighting matrix $\boldsymbol{Q}_{1ij(k)}$. For linear-quadratic control design, positive definiteness of $\boldsymbol{Q}_{2ij(k)}$ is usually required. An "engineering solution" to this problem consists in setting this matrix to a "small" symmetric and positive definite matrix for $\delta_{ij(k)} = 0$. The discretization furthermore generally leads to a non-zero cross weighting matrix $\boldsymbol{Q}_{12ij(k)} \in \mathbb{R}^{(\mathrm{n}_i+\mathrm{m}_i)\times\mathrm{m}_i}$.

Note that if $\tau_{\mathrm{I}i} = 0$ for all control tasks $\mathrm{T}_i \in \mathbb{T}$, then the overall delays become $\tau_i = \tau_{\mathrm{IO}i}$ and the matrices (7.26b) to (7.26d) reduce to

$$\boldsymbol{Q}_{1ij(k)} = \begin{pmatrix} \int\limits_0^{\tau_j(k)} \boldsymbol{\Phi}_{ij(k)}^T(t)\boldsymbol{Q}_{1\mathrm{c}i}\boldsymbol{\Phi}_{ij(k)}(t)dt & \int\limits_0^{\tau_j(k)} \boldsymbol{\Phi}_{ij(k)}^T(t)\boldsymbol{Q}_{1\mathrm{c}i}\boldsymbol{\Gamma}_{1ij(k)}(t)dt \\ * & \int\limits_0^{\tau_j(k)} \boldsymbol{\Gamma}_{1ij(k)}^T(t)\boldsymbol{Q}_{1\mathrm{c}i}\boldsymbol{\Gamma}_{1ij(k)}(t) + \boldsymbol{Q}_{2\mathrm{c}i}dt \end{pmatrix} \tag{7.28a}$$

$$\boldsymbol{Q}_{12ij(k)} = \boldsymbol{0} \tag{7.28b}$$

$$\boldsymbol{Q}_{2ij(k)} = \boldsymbol{0}. \tag{7.28c}$$

An overall cost function can be defined as the sum of the individual cost functions, i.e.

$$J_N = \sum_{i=1}^{M} J_{Ni}. \tag{7.29}$$

Using (7.23a), the overall cost function can be written as

$$J_N = \boldsymbol{x}^T(N)\boldsymbol{Q}_0\boldsymbol{x}(N) + \sum_{k=0}^{N-1}\left[\boldsymbol{x}^T(k)\boldsymbol{Q}_{1j(k)}\boldsymbol{x}(k) + 2\boldsymbol{x}^T(k)\boldsymbol{Q}_{12j(k)}\boldsymbol{u}(k) + \boldsymbol{u}^T(k)\boldsymbol{Q}_{2j(k)}\boldsymbol{u}(k)\right] \tag{7.30}$$

where the block-diagonal weighting matrices $\boldsymbol{Q}_{1j(k)} \in \mathbb{R}^{\mathrm{n}\times\mathrm{n}}$, $\boldsymbol{Q}_{12j(k)} \in \mathbb{R}^{\mathrm{n}\times\mathrm{m}}$, $\boldsymbol{Q}_{2j(k)} \in \mathbb{R}^{\mathrm{m}\times\mathrm{m}}$ are given by

$$\boldsymbol{Q}_0 = \begin{pmatrix} \boldsymbol{Q}_{01} & \cdots & \boldsymbol{0} \\ \vdots & \ddots & \vdots \\ \boldsymbol{0} & \cdots & \boldsymbol{Q}_{0M} \end{pmatrix}, \qquad \boldsymbol{Q}_{1j(k)} = \begin{pmatrix} \boldsymbol{Q}_{11j(k)} & \cdots & \boldsymbol{0} \\ \vdots & \ddots & \vdots \\ \boldsymbol{0} & \cdots & \boldsymbol{Q}_{1Mj(k)} \end{pmatrix} \tag{7.31a}$$

$$\boldsymbol{Q}_{12j(k)} = \begin{pmatrix} \boldsymbol{Q}_{121j(k)} & \cdots & \boldsymbol{0} \\ \vdots & \ddots & \vdots \\ \boldsymbol{0} & \cdots & \boldsymbol{Q}_{12Mj(k)} \end{pmatrix}, \quad \boldsymbol{Q}_{2j(k)} = \begin{pmatrix} \boldsymbol{Q}_{21j(k)} & \cdots & \boldsymbol{0} \\ \vdots & \ddots & \vdots \\ \boldsymbol{0} & \cdots & \boldsymbol{Q}_{2Mj(k)} \end{pmatrix}. \tag{7.31b}$$

The cross weighting matrix $\boldsymbol{Q}_{12j(k)}$ can always be eliminated by transformation of the control vector as shown in [ÅW90, p. 338] or [FPW90, Section 9.4.5] and will therefore not be considered in the following.

8 Control and Scheduling Codesign

8.1 Problem Formulation

The control and scheduling codesign problem can be formulated as

Problem 8.1 *For the NECS (7.21) find a control sequence $\boldsymbol{u}^*(k)$ and a task sequence $j^*(k)$ with $k = 0, \ldots, N-1$ such that the cost function (7.30) is minimized, i.e.*

$$\min_{\substack{\boldsymbol{u}(0),\ldots,\boldsymbol{u}(N-1)\\ j(0),\ldots,j(N-1)}} J_N \qquad \text{subject to (7.21).} \tag{8.1}$$

The NECS (7.21) is essentially a switched system (1.3) where the task sequence is the switching sequence. Problem 8.1 is thus a special case of Problem 1.9. Finite-horizon control and scheduling, receding-horizon control and scheduling and periodic control and scheduling presented in Chapters 2 to 4 can thus be applied to Problem 8.1. However, due to the structural properties of the NECS (7.21) discussed in the following section, some modifications are required.

8.2 Structural Properties of the NECS Model

Proposition 8.2 *The subsystems $(\boldsymbol{A}_{j(k)}, \boldsymbol{B}_{j(k)})$ with $j(k) \in \mathbb{M}$ of the NECS (7.21) are not stabilizable.*

PROOF. To analyze the stabilizability of the pair $(\boldsymbol{A}_{j(k)}, \boldsymbol{B}_{j(k)})$, it is sufficient to analyze the stabilizability of the pair $(\boldsymbol{A}_{ij(k)}, \boldsymbol{B}_{ij(k)})$ for some $i \in \mathbb{M}$ due to the block-diagonal and therefore decoupled structure of the pair $(\boldsymbol{A}_{j(k)}, \boldsymbol{B}_{j(k)})$.

Consider $\delta_{ij(k)} = 0$ for some $i \in \mathbb{M}$, then

$$\boldsymbol{A}_{ij(k)} = \begin{pmatrix} \boldsymbol{\Phi}_{ij(k)}\left(\tau_{j(k)}\right) & \boldsymbol{\Gamma}_{1ij(k)}\left(\tau_{j(k)}, \tau_{j(k)}\right) \\ \boldsymbol{0} & \boldsymbol{I} \end{pmatrix} \tag{8.2a}$$

$$\boldsymbol{B}_{ij(k)} = \begin{pmatrix} \boldsymbol{\Gamma}_{0ij(k)}\left(\tau_{j(k)}, \tau_{j(k)}\right) \\ \boldsymbol{0} \end{pmatrix} = \begin{pmatrix} 0 \\ 0 \end{pmatrix}. \tag{8.2b}$$

Obviously, $\boldsymbol{A}_{ij(k)}$ has eigenvalues $\lambda_r = 1$ with $r \in \{\mathrm{n}_i + 1, \ldots, \mathrm{n}_i + \mathrm{m}_i\}$ which are related to the preceding control vector $\boldsymbol{u}_i(k-1)$ stored in the augmented state vector $\boldsymbol{x}_i(k)$.

These marginally stable eigenvalues are not controllable since $\boldsymbol{B}_{ij(k)} = \boldsymbol{0}$. Consequently, the pair $(\boldsymbol{A}_{ij(k)}, \boldsymbol{B}_{ij(k)})$ and therefore also the pair $(\boldsymbol{A}_{j(k)}, \boldsymbol{B}_{j(k)})$ is not stabilizable. □

Proposition 8.2 has severe consequences for receding-horizon control and scheduling. Both the a priori stability condition given in Theorem 3.3 and the a posteriori stability criteria given in Theorems 3.6 and 3.8 are generally inapplicable for switched systems with non-stabilizable subsystems as outlined in Remarks 3.10 and 3.17. Only the a posteriori stability criterion given in Theorem 3.9 can be utilized as pointed out in Remark 3.19. It is surely desirable to exploit the full repertoire of a posteriori stability criteria. Therefor, a modification of receding-horizon control and scheduling is required which is addressed in the forthcoming chapter.

Proposition 8.2 has also consequences for periodic control and scheduling, in particular for the admissibility of a p-periodic task sequence.

Proposition 8.3 *A p-periodic task sequence $j(k)$ is not admissible if not all control tasks $\mathrm{T}_i \in \mathbb{T}$ are contained, i.e. if $\{j(0)\} \cup \ldots \cup \{j(p-1)\} \neq \mathbb{M}$*

PROOF. In view of Definition 4.9 the stabilizability of the pair $(\overline{\boldsymbol{A}}_{0j}, \overline{\boldsymbol{B}}_{0j})$ and the detectability of the pair $(\overline{\boldsymbol{C}}_{0j}, \overline{\boldsymbol{A}}_{0j})$ must be analyzed. It is, however, sufficient to inspect the stabilizability only.

To analyze the stabilizability of the pair $(\overline{\boldsymbol{A}}_{0j}, \overline{\boldsymbol{B}}_{0j})$, it is sufficient to analyze the stabilizability of the pair $(\overline{\boldsymbol{A}}_{0ij}, \overline{\boldsymbol{B}}_{0ij})$ for some $i \in \mathbb{M}$, following the same rationale as the proof of Proposition 8.2.

Consider $\delta_{ij(k)} = 0$ for some $i \in \mathbb{M}$ and all $k \in \{0, \ldots, p-1\}$, i.e. the control task T_i is not contained in the p-periodic task sequence $j(k)$. The transition matrix of the related periodically time-varying system then has the form

$$\boldsymbol{\Phi}_i(k, 0) = \boldsymbol{A}_{ij(k-1)} \boldsymbol{A}_{ij(k-2)} \cdots \boldsymbol{A}_{ij(0)} = \begin{pmatrix} \bullet & \bullet \\ 0 & \boldsymbol{I} \end{pmatrix} \quad \text{for } k > 0, \tag{8.3}$$

leading to a lifted system and input matrix having the form

$$\overline{\boldsymbol{A}}_{0ij} = \boldsymbol{\Psi}_i(0) = \begin{pmatrix} \bullet & \bullet \\ 0 & \boldsymbol{I} \end{pmatrix} \tag{8.4a}$$

$$\begin{aligned} \overline{\boldsymbol{B}}_{0ij} &= \begin{pmatrix} \boldsymbol{\Phi}_i(0+p, 0+1)\check{\boldsymbol{B}}_{ij(0)} & \boldsymbol{\Phi}_i(0+p, 0+2)\check{\boldsymbol{B}}_{ij(0+1)} & \cdots & \check{\boldsymbol{B}}_{ij(0+p-1)} \end{pmatrix} \\ &= \begin{pmatrix} 0 & \ldots & 0 \\ 0 & \ldots & 0 \end{pmatrix}. \end{aligned} \tag{8.4b}$$

Using the argumentation from the proof of Proposition 8.2, it can be concluded that the pair $(\overline{\boldsymbol{A}}_{ij(0)}, \overline{\boldsymbol{B}}_{ij(0)})$ and therefore also the pair $(\overline{\boldsymbol{A}}_{ij(0)}, \overline{\boldsymbol{B}}_{ij(0)})$ is not stabilizable. □

Remark 8.1. Proposition 8.3 provides a necessary condition for admissibility, i.e. a p-periodic task sequence $j(k)$ can only be admissible if all control tasks $\mathrm{T}_i \in \mathbb{T}$ are contained. As a direct consequence, a p-periodic task sequence $j(k)$ can only be admissible if $p \geq M$.

Remark 8.2. If a ZOH behavior according to (7.20) is utilized and all plants $\mathrm{P}_i \in \mathbb{P}$ are stable, then all subsystems $(\boldsymbol{A}_{j(k)}, \boldsymbol{B}_{j(k)})$, $j(k) \in \mathbb{M}$ of the NECS (7.21) are stabilizable. Furthermore, a p-periodic task sequence $j(k)$ can then be admissible also if not all control tasks $\mathrm{T}_i \in \mathbb{T}$ are contained. This can be easily seen following the lines of the proofs of Propositions 8.2 and 8.3. A ZOH behavior according to (7.20) is, however, not always appropriate as outlined in Remark 7.4. Furthermore, sometimes not all plants $\mathrm{P}_i \in \mathbb{P}$ are stable.

9 N'-Step Receding-Horizon Control and Scheduling

The subsystems of the NECS (7.21) can not be stabilized individually as indicated by Proposition 8.2 but may be stabilized by switching as suggested by Proposition 8.3. Based on this observation, receding-horizon control and scheduling can be modified such that Theorems 3.6 and 3.8 are applicable for an a posteriori stability analysis. The idea consists in applying not the first element but the first N' elements of the control and task sequence to the NECS and thus in reoptimizing not at each time instant but each N' time instants, leading to an N'-step receding-horizon control and scheduling strategy. The resulting closed-loop system can be lifted over N' time instants. The stability of the lifted closed-loop system can then be analyzed based on Theorems 3.6 and 3.8.

9.1 Problem Formulation

For formalizing N'-step receding-horizon control and scheduling, the notion of N'-step admissibility of a task sequence is crucial.

Definition 9.1 *A task sequence $j(k), \dots, j(k+N-1)$ is called N'-step admissible if the first N' elements $j(k), \dots, j(k+N'-1)$ contain all control tasks $\mathrm{T}_i \in \mathbb{T}$, i.e. if $\{j(k)\} \cup \ldots \cup \{j(k+N'-1)\} = \mathbb{M}$, where $M \leq N' \leq N$.*

The N'-step admissibility of a task sequence is a necessary condition for the applicability of Theorems 3.6 and 3.8 as will become clear from the subsequent considerations.

Problem 9.2 *For the NECS (7.21) and the current state $\boldsymbol{x}(k)$ find at time instants $k = 0, N', 2N', \dots$ a control sequence $\boldsymbol{u}^*(k), \dots, \boldsymbol{u}^*(k+N-1)$ and an N'-step admissible task sequence $j^*(k), \dots, j^*(k+N-1)$ such that the cost function (7.30) reformulated according to (3.1) is minimized over the prediction horizon N, i.e.*

$$\begin{aligned} &\min_{\substack{\boldsymbol{u}(k),\dots,\boldsymbol{u}(k+N-1)\\ j(k),\dots,j(k+N-1)}} J'_N(k) \\ &\text{subject to } \boldsymbol{x}(k+1+i) = \boldsymbol{A}_{j(k+i)}\boldsymbol{x}(k+i) + \boldsymbol{B}_{j(k+i)}\boldsymbol{u}(k+i),\ i = 0, \dots, N-1. \end{aligned} \tag{9.1}$$

If the first N' elements $\boldsymbol{u}^*(k), \dots, \boldsymbol{u}^*(k+N'-1)$ of the control sequence and the first N' elements $j^*(k), \dots, j^*(k+N'-1)$ of the task sequence are then applied to the NECS, an

N'-step receding-horizon control and scheduling (N'-step RHCS) strategy is obtained.

Remark 9.1. Optimality can be relaxed in Problem 9.2 analogously to Problem 2.7, yielding a relaxed N'-step receding-horizon control and scheduling (N'-step RRHCS) strategy.

Problem 9.2 evidently parallels Problem 3.1. The ideas presented in Chapters 2 and 3 on finite- and receding-horizon control and scheduling can therefore be extended for N'-step receding-horizon control and scheduling. In the following sections this extension is described.

9.2 Dynamic Programming Solution

The dynamic programming solution formulated in Theorem 2.2 together with pruning (Theorems 2.4, 2.5 and 2.6) and relaxation (Theorems 2.8 and 2.9) is also applicable to Problem 9.2. Task sequences which are not N'-step admissible must, however, be removed.

The N'-step admissibility of a task sequence $j(0), \ldots, j(N-1)$ can be detected during dynamic programming. Consider step k' in the Bellman equation (2.11) or the relaxed Bellman equation (2.33). The subsequence $j(k'), \ldots, j(N-1)$ is then already fixed. For N'-step admissibility only the subsequence $j(k'), \ldots, j(N'-1)$ is relevant. The number of tasks contained in this subsequence is given by

$$M_{\text{cont}}(k') = |\{j(k')\} \cup \ldots \cup \{j(N'-1)\}| \,. \tag{9.2}$$

The subsequence $j(0), \ldots, j(k'-1)$ is not yet fixed, i.e. k' elements can still be selected to render the overall task sequence N'-step admissible, see Figure 9.1.

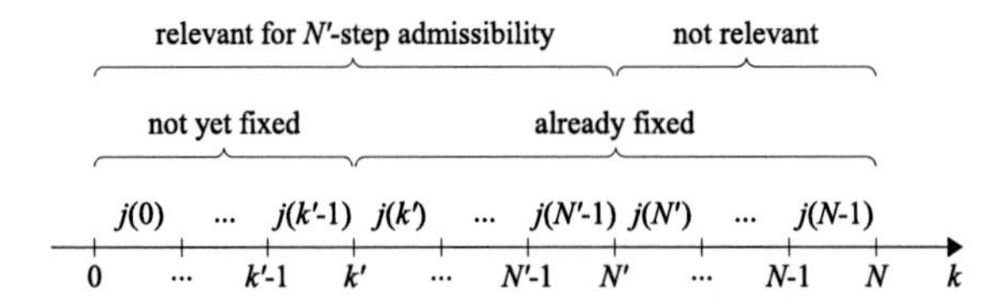

Figure 9.1: Illustration of the N'-step admissibility of a task sequence $j(0), \ldots, j(N-1)$

The overall task sequence can, however, not be rendered N'-step admissible if the number of tasks not yet contained which is given by $M - M_{\text{cont}}(k')$ is larger than the number of remaining elements k'. This leads to

Proposition 9.3 (Removal Criterion) *A task sequence $j(0), \ldots, j(N-1)$ is not N'-step admissible and can thus be marked removable at step k' in the Bellman equation (2.11)*

or the relaxed Bellman equation (2.33) if

$$k' < M - M_{\mathrm{cont}}(k'). \tag{9.3}$$

Remark 9.2. Since $M_{\mathrm{cont}}(k') \geq 1$ for all $k' \in \{0, \ldots, N'-1\}$, the removal criterion must only be evaluated at steps $k' < M - 1$.

Remark 9.3. The removal of task sequences which are not N'-step admissible clearly induces suboptimality. Therefore, methods for a suboptimality analysis are required. This is the subject of the following section.

Remark 9.4. The parameter N' can be considered as a design parameter. The choice of N' can be based on the following considerations:

Let $\mathbb{J}_N$ be the set of task sequences and $\mathbb{J}'_N$ be the set of N'-step admissible task sequences for the time horizon N. The number of task sequences $|\mathbb{J}_N|$ is determined by $|\mathbb{J}_N| = M^N$. The number of N'-step admissible task sequences $|\mathbb{J}'_N|$ can be obtained by evaluating the removal criterion given in Proposition 9.3. In Figure 9.2, $|\mathbb{J}_N|$, $|\mathbb{J}'_N|$ and their ratio $|\mathbb{J}'_N|/|\mathbb{J}_N|$ are depicted for different M, N and N'.

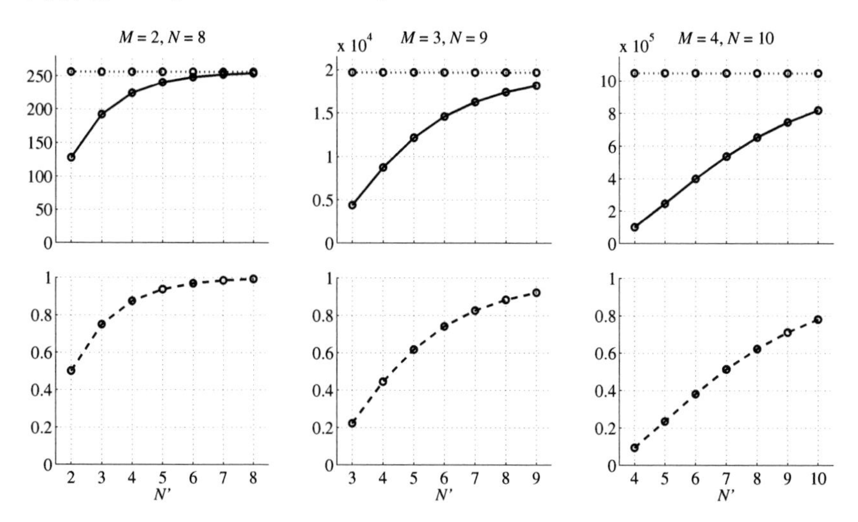

Figure 9.2: $|\mathbb{J}_N|$ (dotted), $|\mathbb{J}'_N|$ (solid) and $|\mathbb{J}'_N|/|\mathbb{J}_N|$ (dashed) for different M, N and N'

$|\mathbb{J}'_N|$ and $|\mathbb{J}'_N|/|\mathbb{J}_N|$ increase degressively with N'. Hence, the suboptimality induced by removing task sequences which are not N'-step admissible can be reduced by enlarging N'. On the other hand, for larger N' more steps of the control and task sequence have to be applied between reoptimizations whereby the adaptivity under disturbances is reduced. Note that for larger M the ratio $|\mathbb{J}'_N|/|\mathbb{J}_N|$ generally reduces.

9.3 Suboptimality Analysis

Methods for a suboptimality analysis can be deduced from concepts for relaxation. Specifically, Theorems 2.8 and 2.9 can be reconsidered for a suboptimality analysis.

The suboptimality analysis problem can be formally stated as

Problem 9.4 *Find the minimum suboptimality factor $\alpha' \in \mathbb{R}$ with $\alpha' \geq 1$ for which*

$$J_N^*(k) \leq J_N'^*(k) \leq \alpha' J_N^*(k) \tag{9.4}$$

where $J_N^*(k) = \min_{l \in \mathbb{L}_0} \boldsymbol{x}^T(k)\boldsymbol{P}_l\boldsymbol{x}(k)$ and $J_N'^*(k) = \min_{l' \in \mathbb{L}_0'} \boldsymbol{x}^T(k)\boldsymbol{P}_{l'}\boldsymbol{x}(k)$ are the minimum costs before and after removal.

This problem can be approached based on

Theorem 9.5 (Suboptimality Criterion) *If for each $\boldsymbol{P}_m$, $m \in \bar{\mathbb{L}}_0' = \mathbb{L}_0 \setminus \mathbb{L}_0'$ (associated to a removed task sequence) and each $\boldsymbol{x}(k) \in \mathbb{R}^n$ there exists at least one $\boldsymbol{P}_{l'}$, $l' \in \mathbb{L}_0'$ (associated to a retained task sequence) such that*

$$\boldsymbol{x}^T(k)\alpha'\boldsymbol{P}_m\boldsymbol{x}(k) \geq \boldsymbol{x}^T(k)\boldsymbol{P}_{l'}\boldsymbol{x}(k), \tag{9.5}$$

then inequality (9.4) is satisfied.

Proof. By construction. Note that since $\mathbb{L}_0' \subseteq \mathbb{L}_0$ the inequality (9.5) is fulfilled trivially by $l' = m$ for all $m \in \mathbb{L}_0'$ and arbitrary $\alpha' \geq 1$. Thus, only $m \in \bar{\mathbb{L}}_0' = \mathbb{L}_0 \setminus \mathbb{L}_0'$ must be inspected. □

Remark 9.5. The suboptimality criterion can not be evaluated exactly due to the dependence of inequality (9.5) on $\boldsymbol{x}(k) \in \mathbb{R}^n$. Instead, a common controller index $l' = n = \text{const.}$ for all states $\boldsymbol{x}(k) \in \mathbb{R}^n$ or a convex combination of $\boldsymbol{P}_{l'}$ exploiting the S-procedure can be used analogously to Theorems 2.5 and 2.6. The suboptimality analysis then yields an upper bound on the suboptimality factor α'. For brevity of presentation only the latter approach will be considered in the following.

Remark 9.6. The suboptimality analysis can be combined with relaxed pruning since some $\boldsymbol{P}_{l'}$, $l' \in \mathbb{L}_0'$ may be dispensable for fulfillment of the suboptimality criterion for a given α' and may thus be pruned. Thereby, the complexity can be further reduced.

For evaluating the suboptimality criterion, an efficient domination criterion can be formulated based on the S-procedure as

Theorem 9.6 (Domination Criterion) *If there exist non-negative scalars $\xi_{l'}$ and $\boldsymbol{P}_{l'}$ with $l' \in \mathbb{L}_0'$ such that*

$$\sum_{l' \in \mathbb{L}_0'} \xi_{l'} = 1 \quad \text{and} \quad \alpha'\boldsymbol{P}_m \succeq \sum_{l' \in \mathbb{L}_0'} \xi_{l'}\boldsymbol{P}_{l'}, \tag{9.6}$$

then $\alpha'\boldsymbol{P}_m$ with $m \in \bar{\mathbb{L}}_0'$ is dominated.

PROOF. Follows the same lines as the proof of Theorem 2.6. □

An algorithm for determining an upper bound on the suboptimality factor α' utilizing Theorem 9.6 and Remark 9.6 is given in Algorithm 9.1.

Algorithm 9.1 Suboptimality Analysis

Input: $\boldsymbol{P}_l$, $l \in \mathbb{L}_0$ obtained before removal, $\boldsymbol{P}_{l'}$, $l' \in \mathbb{L}'_0$ obtained after removal,
$\underline{\alpha}'$ not fulfilling Theorem 9.5, $\overline{\alpha}'$ fulfilling Theorem 9.5, accuracy ε

Output: $\boldsymbol{P}_{\hat{l}'}$, $\hat{l}' \in \hat{\mathbb{L}}'_0$ after relaxed pruning, suboptimality factor α'

while $\frac{1}{2}(\overline{\alpha}' - \underline{\alpha}') > \varepsilon$ **do**
  $\alpha'_{\text{mid}} = \frac{1}{2}(\underline{\alpha}' + \overline{\alpha}')$ // *determine midpoint*
  $\hat{\mathbb{L}}'_0 = \{\}$ // *initialize relaxed controller index set* $\hat{\mathbb{L}}'_0$
  fulfilled = true // *suboptimality criterion fulfilled (initialization)*
  for each $m \in \mathbb{L}'_0$ **do**
    if $\nexists\, \xi_{\hat{l}'} \geq 0, \hat{l}' \in \hat{\mathbb{L}}'_0$: $\sum_{\hat{l}'} \xi_{\hat{l}'} = 1$, $\alpha'_{\text{mid}} \boldsymbol{P}_m \succeq \sum_{\hat{l}'} \xi_{\hat{l}'} \boldsymbol{P}_{\hat{l}'}$ **then**
      $\hat{\mathbb{L}}'_0 = \hat{\mathbb{L}}'_0 \cup \{m\}$ // *add controller index*
    end if
  end for
  for each $m \in \bar{\mathbb{L}}'_0$ **do**
    if $\nexists\, \xi_{\hat{l}'} \geq 0, \hat{l}' \in \hat{\mathbb{L}}'_0$: $\sum_{\hat{l}'} \xi_{\hat{l}'} = 1$, $\alpha'_{\text{mid}} \boldsymbol{P}_m \succeq \sum_{\hat{l}'} \xi_{\hat{l}'} \boldsymbol{P}_{\hat{l}'}$ **then**
      fulfilled = false // *suboptimality criterion not fulfilled*
      break for loop
    end if
  end for
  if fulfilled = true **then**
    $\overline{\alpha}' = \alpha'_{\text{mid}}$
  else
    $\underline{\alpha}' = \alpha'_{\text{mid}}$
  end if
end while
$\alpha' = \frac{1}{2}(\overline{\alpha}' + \underline{\alpha}')$

The algorithm is based on the bisection method:

The suboptimality factor α' is always contained in an interval $[\underline{\alpha}', \overline{\alpha}']$ with the upper endpoint $\overline{\alpha}'$ fulfilling the suboptimality criterion and the lower endpoint $\underline{\alpha}'$ not fulfilling the suboptimality criterion. This interval is bisected by the midpoint $\alpha'_{\text{mid}} = \frac{1}{2}(\underline{\alpha}' + \overline{\alpha}')$. The midpoint α'_{mid} is analyzed for fulfillment of the suboptimality criterion. In case of fulfillment, the upper endpoint $\overline{\alpha}'$ is set to α'_{mid}. Otherwise, the lower endpoint $\underline{\alpha}'$ is set to α'_{mid}. This bisection is iterated until the suboptimality factor $\alpha' = \frac{1}{2}(\underline{\alpha}' + \overline{\alpha}')$ is determined with a predefined accuracy $\frac{1}{2}(\overline{\alpha}' - \underline{\alpha}') \leq \varepsilon$. The convergence of the bisection is guaranteed. The accuracy after n iterations is given by

$$\varepsilon(n) = \frac{\overline{\alpha}' - \underline{\alpha}'}{2^{n+1}} \tag{9.7}$$

where $\overline{\alpha}'$ and $\underline{\alpha}'$ are the initial endpoints. For a predefined accuracy ε, the required number of iterations follows by rearranging (9.7) as

$$n(\varepsilon) = \left\lceil \log_2 \left(\frac{\overline{\alpha}' - \underline{\alpha}'}{\varepsilon} \right) - 1 \right\rceil . \tag{9.8}$$

A detailed discussion on the bisection method can be found e.g. in [CK08, Section 3.1]. An illustration of the bisection method is given in Figure 9.3.

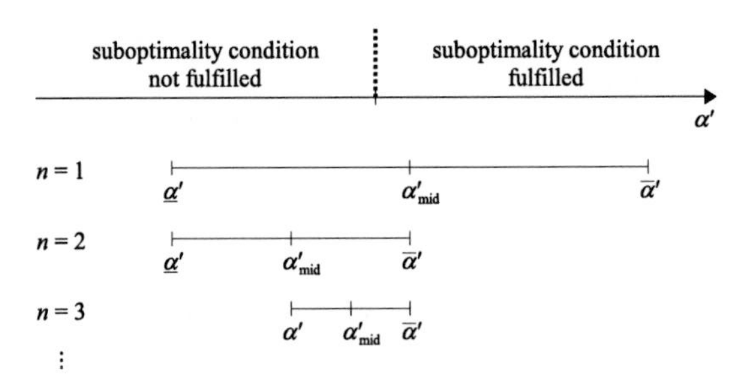

Figure 9.3: Illustration of the bisection method

Within the bisection procedure, the fulfillment of the suboptimality criterion is addressed in two steps:

First, relaxed pruning is applied for the current α'_{mid} to the DRE solutions $\boldsymbol{P}_{l'}$, $l' \in \mathbb{L}'_0$ obtained after removal, leading to the relaxed set of controller indices $\hat{\mathbb{L}}'_0$. Then, the fulfillment of the suboptimality criterion is analyzed by checking the domination criterion for each DRE solution $\boldsymbol{P}_m$, $m \in \bar{\mathbb{L}}'_0$ associated to an N'-step non-admissible task sequence. If a single $\boldsymbol{P}_m$ does not fulfill the domination criterion, then checking can be stopped prematurely. To increase the efficiency of relaxed pruning and detect the non-fulfillment of the domination criterion early, $\boldsymbol{P}_{l'}$, $l' \in \mathbb{L}'_0$ and $\boldsymbol{P}_{\bar{l}'}$, $\bar{l}' \in \bar{\mathbb{L}}'_0$ can be presorted by trace as outlined in Section 2.4.

These two steps together with the preceding removal are depicted in Figure 9.4 from a set partitioning perspective.

Remark 9.7. If relaxed dynamic programming with a relaxation factor α is utilized to solve Problem 9.2, then the total suboptimality factor is given by $\alpha'_{\text{tot}} = \alpha\alpha'$.

Remark 9.8. The suboptimality analysis can alternatively be incorporated into relaxed dynamic programming. The idea is to increase the relaxation factor during the backward

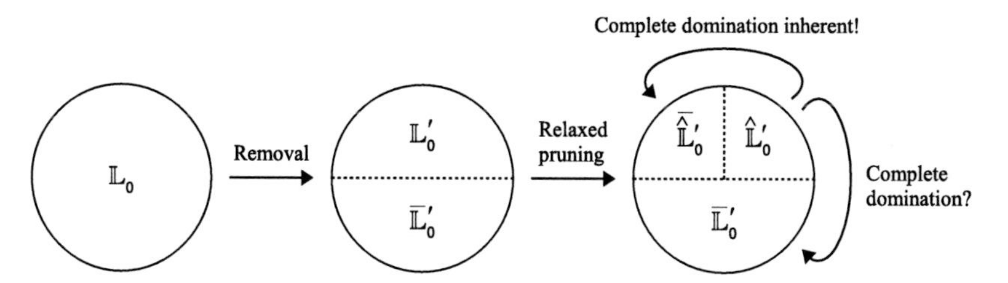

Figure 9.4: Illustration of removal, relaxed pruning and domination in Algorithm 9.1

iteration of the relaxed Bellman equation such that in each step only N'-step admissible task sequences remain. For identifying N'-step admissible task sequences, the removal criterion given in Theorem 9.3 can be utilized. In numerical studies this approach has however proved very conservative in terms of the obtained upper bound on α' when comparing to separated relaxed dynamic programming and suboptimality analysis.

Remark 9.9. The suboptimality analysis can be analogously applied for the $\mathrm{HPCS_{on}}$ and the OPP strategy presented in Section 4.2.4.

Remark 9.10. Note that the suboptimality factor α specifies the suboptimality w.r.t. the finite-horizon cost $J_N^*(k)$ according to (3.1) but not w.r.t. the infinite horizon cost J_∞ according to (2.1). To specify this suboptimality, an additional suboptimality analysis as outlined in Remark 3.2 is required.

9.4 Explicit Solution

Theorem 9.7 *The solution to Problem 9.2 is given by the piecewise linear (PWL) state feedback control law*

$$\boldsymbol{u}^*(k) = -\boldsymbol{K}_{m(k \bmod N')}\boldsymbol{x}(k) \quad for \quad \boldsymbol{x}(\lfloor k/N' \rfloor N') \in \mathcal{X}_{m(0)} \tag{9.9}$$

where the regions $\mathcal{X}_{m(0)}$ *with* $\bigcup_{l'(0)\in\mathbb{L}_0'} \mathcal{X}_{l'(0)} = \mathbb{R}^n$ *are described by*

$$\mathcal{X}_{m(0)} = \left\{\boldsymbol{x}(k) \mid \boldsymbol{x}^T(k)\boldsymbol{P}_{m(0)}\boldsymbol{x}(k) \leq \boldsymbol{x}^T(k)\boldsymbol{P}_{l'(0)}\boldsymbol{x}(k) \; \forall l'(0); m(0), l'(0) \in \mathbb{L}_0'\right\}. \tag{9.10}$$

PROOF. The proof follows the same lines as the proof of Theorem 3.2. Under the N'-step RHCS strategy, however, the region membership test is performed only at time instants $k = 0, N', 2N', \ldots$, leading to the optimal controller index $m(0) \in \mathbb{L}_0'$. The subsequent optimal controller indices $m(i) \in \mathbb{L}_i'$ with $i = 1, \ldots, N'-1$ follow from iterating the switching tree backwards beginning with $m(0)$. The optimal switching indices result from $v(i) = \sigma(m(i))$. An illustration of the N'-step RHCS strategy is given in Figure 9.5. □

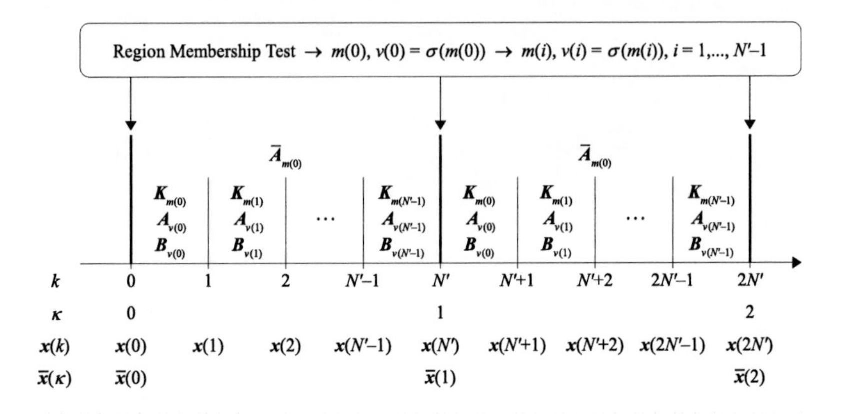

Figure 9.5: Illustration of the N'-step RHCS strategy and the lifting procedure

Remark 9.11. The PWL state feedback control law (9.9) is *explicit*. The feedback matrices $\boldsymbol{K}_{l'(i)}$, $i = 0, \dots, N'-1$ and the DRE solutions $\boldsymbol{P}_{l'(0)}$ can be calculated offline for all $l'(0) \in \mathbb{L}'_0$ and stored. The online computation reduces to the evaluation of the PWL state feedback control law (9.9).

Remark 9.12. The statements on the well-posedness of the solution given in Remark 3.4 and on the non-convexity of the regions given in Remark 3.5 also apply to Theorem 9.7.

Remark 9.13. Substituting (9.9) into (7.21) leads to the autonomous PWL closed-loop system

$$\boldsymbol{x}(k+1) = \widetilde{\boldsymbol{A}}_{m(k \bmod N')}\boldsymbol{x}(k) \quad \text{for } \boldsymbol{x}(\lfloor k/N' \rfloor N') \in \mathcal{X}_{m(0)}, \quad \boldsymbol{x}(0) = \boldsymbol{x}_0 \tag{9.11}$$

with $\widetilde{\boldsymbol{A}}_{m(i)} = \boldsymbol{A}_{v(i)} - \boldsymbol{B}_{v(i)}\boldsymbol{K}_{m(i)}$, $i = 0, \dots, N'-1$ where the task index $v(i) \in \mathbb{M}$ follows from $v(i) = \sigma(m(i))$ for a given region $\mathcal{X}_{m(0)}$ with $m(0) \in \mathbb{L}'_0$.

Remark 9.14. The explicit PWL state feedback control law (9.9) and the PWL closed-loop system (9.11) can also be formulated in terms of the N'-step RRHCS strategy using $\hat{l}'(0) \in \hat{\mathbb{L}}'_0$ instead of $l'(0) \in \mathbb{L}'_0$.

Remark 9.15. The PWL closed-loop system (9.11) can be lifted over N' time instants. The concept of lifting has already been introduced in Section 4.1.2 for periodic control. The state vector $\boldsymbol{x}(k)$ is downsampled over N' time instants, leading to the lifted state vector

$$\overline{\boldsymbol{x}}(\kappa) = \boldsymbol{x}(N'\kappa) \tag{9.12}$$

with $\overline{\boldsymbol{x}}(\kappa) \in \mathbb{R}^n$ and $\kappa = \lfloor k/N' \rfloor$ denoting the time instant of the lifted system. The lifted system matrix results from solving the PWL closed-loop system (9.11) over N' time instants as

$$\overline{\boldsymbol{A}}_{m(0)} = \widetilde{\boldsymbol{A}}_{m(N'-1)}\widetilde{\boldsymbol{A}}_{m(N'-2)} \cdots \widetilde{\boldsymbol{A}}_{m(0)} \tag{9.13}$$

with $\overline{\boldsymbol{A}}_{m(0)} \in \mathbb{R}^{n \times n}$. The lifted PWL closed-loop system is then represented by

$$\overline{\boldsymbol{x}}(\kappa+1) = \overline{\boldsymbol{A}}_{m(0)}\overline{\boldsymbol{x}}(\kappa) \quad \text{for} \quad \overline{\boldsymbol{x}}(\kappa) \in \mathcal{X}_{m(0)}, \quad \overline{\boldsymbol{x}}(0) = \boldsymbol{x}_0. \tag{9.14}$$

The lifting procedure is illustrated in Figure 9.5.

9.5 Stability Analysis

The PWL closed-loop system (9.11) is stable evidently iff the lifted PWL closed-loop system (9.14) is stable. The stability analysis can therefore focus on the latter.

The stability properties of the lifted PWL closed-loop system (9.14) are strongly related to N'-step admissibility:

Proposition 9.8 *The subsystem $\overline{\boldsymbol{A}}_{m(0)}$ with $m(0) \in \mathbb{L}_0'$ of the lifted PWL closed-loop system (9.14) is not globally asymptotically stable if the associated task sequence $v(i) = \sigma(m(i))$ with $i = 0, \ldots, N-1$ is not N'-step admissible.*

PROOF. The proof follows similar arguments as the proofs of Propositions 8.2 and 8.3. To analyze the stability of the subsystem $\overline{\boldsymbol{A}}_{m(0)}$, it is sufficient to analyze the stability of the subsystem $\overline{\boldsymbol{A}}_{\iota m(0)}$ associated to task T_ι, $\iota \in \mathbb{M}$ owing to the block-diagonal and therefore decoupled structure of the NECS (7.21).

Assume that the task sequence $v(i) = \sigma(m(i))$ with $i = 0, \ldots, N-1$ is not N'-step admissible. This implies $\delta_{\iota m(i)} = 0$ and consequently $\boldsymbol{B}_{\iota m(i)} = \boldsymbol{0}$ for some $\iota \in \mathbb{M}$ and all $i \in \{0, \ldots, N'-1\}$, leading to

$$\overline{\boldsymbol{A}}_{\iota m(0)} = \boldsymbol{A}_{\iota m(N'-1)}\boldsymbol{A}_{\iota m(N'-2)} \cdots \boldsymbol{A}_{\iota m(0)} = \begin{pmatrix} \bullet & \bullet \\ 0 & \boldsymbol{I} \end{pmatrix}. \tag{9.15}$$

From the argumentation of the proof of Proposition 8.2, it follows that the subsystem $\overline{\boldsymbol{A}}_{\iota m(0)}$ and therefore the subsystem $\overline{\boldsymbol{A}}_{m(0)}$ is marginally stable or unstable. □

Proposition 9.8 together with Remark 3.17 entails that the N'-step admissibility of all task sequences is a necessary condition for the applicability of Theorems 3.6 and 3.8 for analyzing the stability of the lifted PWL closed-loop system (9.14). The statement made at the beginning of this chapter is thus verified. However, the N'-step admissibility of all task sequences is not a sufficient condition. Therefore, some subsystems $\overline{\boldsymbol{A}}_{m(0)}$ may still be marginally stable or unstable. This problem can be addressed by two approaches:

First, the marginally stable and unstable subsystems $\overline{\boldsymbol{A}}_{m(0)}$ can be removed. The sub-optimality analysis must then be reperformed.

Second, the terminal weighting matrix $\boldsymbol{Q}_0$, the prediction horizon N or the prediction subhorizon N' can be increased, reconsidering the discussion at the end of Section 3.3.2.

An initial guess for the terminal weighting matrix $\boldsymbol{Q}_0$ can in view of Remark 3.10 however not be obtained from Theorems 3.3 and 3.4. N'-step receding-horizon control and scheduling is clearly related to periodic control and scheduling. Differences primarily concern the constraints on the task sequence. Thus, a reasonable initial guess for $\boldsymbol{Q}_0$ can be obtained based on the PCS$_{\text{off}}$ strategy, i.e. $\boldsymbol{Q}_0 = \boldsymbol{P}_{j^*}(0)$ where $\boldsymbol{P}_{j^*}(0)$ follows from Theorem 4.11 using the expected value of the cost (4.37) or the maximum value of the cost (4.38) and e.g. $p = N'$.

In summary, the subsystems $\overline{\boldsymbol{A}}_{m(0)}$ can always be rendered globally asymptotically stable by imposing N'-step admissibility of all tasks sequences and utilizing the approaches proposed above, making Theorems 3.6 and 3.8 applicable.

Employing Theorems 3.6 and 3.8 for analyzing the stability of the lifted PWL closed-loop system (9.14) necessitates a reformulation:

Theorem 9.9 (A Posteriori Stability Criterion) *If there exist symmetric matrices $\boldsymbol{S}_{m(0)}$ with $m(0) \in \mathbb{L}_0'$ such that the LMIs*

$$\boldsymbol{S}_{m(0)} \succ \boldsymbol{0} \tag{9.16a}$$

$$\overline{\boldsymbol{A}}_{m(0)}^T \boldsymbol{S}_{n(0)} \overline{\boldsymbol{A}}_{m(0)} - \boldsymbol{S}_{m(0)} \prec \boldsymbol{0} \tag{9.16b}$$

are feasible for all pairs of consecutive controller indices $(m(0), n(0)) \in \mathbb{L}_0'^2$, then the lifted PWL closed-loop system (9.14) is globally uniformly asymptotically stable.

Theorem 9.10 (A Posteriori Stability Criterion Based on the S-Procedure) *If there exist symmetric matrices $\boldsymbol{S}_{m(0)}$ with $m(0) \in \mathbb{L}_0'$ and non-negative scalars $\alpha_{m(0)l'(0)}$, $\beta_{m(0)n(0)l'(0)}$ and $\gamma_{m(0)n(0)l'(0)}$ with $m(0), n(0), l'(0) \in \mathbb{L}_0'$ such that the LMIs*

$$\boldsymbol{S}_{m(0)} - \sum_{l'(0)\in\mathbb{L}_0'} \alpha_{m(0)l'(0)} (\boldsymbol{P}_{l'(0)} - \boldsymbol{P}_{m(0)}) \succ \boldsymbol{0} \tag{9.17a}$$

$$\overline{\boldsymbol{A}}_{m(0)}^T \boldsymbol{S}_{n(0)} \overline{\boldsymbol{A}}_{m(0)} - \boldsymbol{S}_{m(0)} + \underset{m(0),n(0)}{\boldsymbol{\Delta}} (\beta_{m(0)n(0)l'(0)}, \gamma_{m(0)n(0)l'(0)}) \prec \boldsymbol{0} \tag{9.17b}$$

with

$$\begin{aligned} &\underset{m(0),n(0)}{\boldsymbol{\Delta}} (\beta_{m(0)n(0)l'(0)}, \gamma_{m(0)n(0)l'(0)}) \\ &= \sum_{l'(0)\in\mathbb{L}_0'} \left[\beta_{m(0)n(0)l'(0)} (\boldsymbol{P}_{l'(0)} - \boldsymbol{P}_{m(0)}) + \gamma_{m(0)n(0)l'(0)} \overline{\boldsymbol{A}}_{m(0)}^T (\boldsymbol{P}_{l'(0)} - \boldsymbol{P}_{n(0)}) \overline{\boldsymbol{A}}_{m(0)} \right]. \end{aligned} \tag{9.18}$$

are feasible for all pairs of consecutive controller indices $(m(0), n(0)) \in \mathbb{L}_0'^2$, then the lifted PWL closed-loop system (9.14) is globally uniformly asymptotically stable.

Remark 9.16. The Remarks 3.13 and 3.15 on the complexity and the Remark 3.16 on reducing the conservatism by regarding the reachability analogously apply to Theorems 9.9 and 9.10.

9.6 Reachability Analysis

The reachability between two regions $\mathcal{X}_{m(0)}$ and $\mathcal{X}_{n(0)}$ can be examined also for the lifted PWL closed-loop system (9.14). The non-reachability can be defined by

Definition 9.11 (Non-Reachability) *Region $\mathcal{X}_{n(0)}$ is not reachable from region $\mathcal{X}_{m(0)}$ (i.e. $\mathcal{X}_{m(0)} \nrightarrow \mathcal{X}_{n(0)}$) iff $\boldsymbol{x}(\kappa+1) = \overline{\boldsymbol{A}}_{m(0)}\boldsymbol{x}(\kappa) \notin \mathcal{X}_{n(0)}$ for all $\boldsymbol{x}(\kappa) \in \mathcal{X}_{m(0)}$. Otherwise region $\mathcal{X}_{n(0)}$ is reachable from region $\mathcal{X}_{m(0)}$ (i.e. $\mathcal{X}_{m(0)} \rightarrow \mathcal{X}_{n(0)}$).*

The non-reachability criterion can then be formulated as

Theorem 9.12 (Non-Reachability Criterion) *If the inequality*

$$\underset{m(0),n(0)}{\boldsymbol{\Delta}}(\beta_{l'(0)}, \gamma_{l'(0)}) \prec \boldsymbol{0} \tag{9.19}$$

holds for some non-negative scalars $\beta_{l'(0)}, \gamma_{l'(0)}$ with $l'(0) \in \mathbb{L}'_0$, then $\mathcal{X}_{m(0)} \nrightarrow \mathcal{X}_{n(0)}$.

Remark 9.17. The Remark 3.21 on the complexity also holds for Theorem 9.12. Furthermore, the reachability matrix $\boldsymbol{R}$ and the reachability ratio ρ can be defined similar as in Remark 3.22.

Remark 9.18. The controller indices $m(i)$, $i = 0, \ldots, N'-1$ within the prediction subhorizon N' are fixed at time instants $\lfloor k/N' \rfloor N'$. A reachability analysis within the prediction subhorizon N' is therefore not required.

10 Case Study

10.1 Introduction

The inverted pendulum is a classic control problem which is widely used as a benchmark for evaluating control strategies. The schematic of an inverted pendulum on a cart is shown in Figure 10.1.

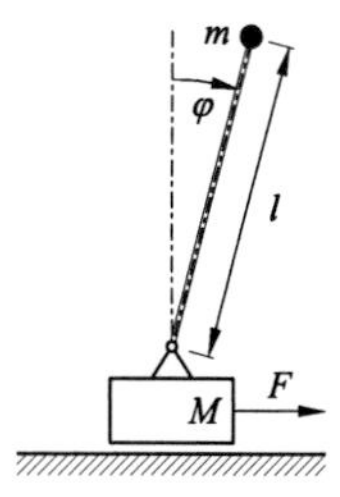

Figure 10.1: Inverted pendulum on a cart

The objective is to maintain the pendulum in the upright position $\varphi = 0$ by exerting a force F on the cart. The inverted pendulum is related to various technical systems, e.g. personal transporters like the Segway®, rocket guidance systems and magnetic bearing systems, and is therefore also of practical relevance.

In the following sections networked control of three inverted pendulums is studied to evaluate the control and scheduling codesign strategies proposed in Chapters 8 and 9.

10.2 Modeling

10.2.1 Computation and Communication Model

Consider a set of three inverted pendulums $\mathbb{P} = \{\mathrm{P}_i, i \in \{1,2,3\}\}$ which are controlled by a set of control tasks $\mathbb{T} = \{\mathrm{T}_i, i \in \{1,2,3\}\}$. The tasks are implemented on a single processor with high performance. Due the high performance the computation times are negligible, i.e. $\tau_{\mathrm{S}} = \tau_{\mathrm{C}i} = 0$. Furthermore, the control signals are transmitted over a network with transmission times $\tau_{\mathrm{CA}i} = 3.5\,\mathrm{ms}$ while the sensor signals are not transmitted over a network, i.e. $\tau_{\mathrm{SC}i} = 0$. No idling is considered, i.e. $\tau_{\mathrm{I}i} = 0$. The input-output delay and the overall delay are then given by $\tau_i = \tau_{\mathrm{IO}i} = 3.5\,\mathrm{ms}$.

10.2.2 Pendulum Model

Each inverted pendulum $\mathrm{P}_i \in \mathbb{P}$ is described by the linearized continuous-time state equation

$$\underbrace{\begin{pmatrix} \dot{\varphi}_i(t) \\ \ddot{\varphi}_i(t) \end{pmatrix}}_{\dot{\boldsymbol{x}}_{\mathrm{c}i}(t)} = \underbrace{\begin{pmatrix} 0 & 1 \\ \frac{m_i+M_i}{M_i}\frac{g}{l_i} & 0 \end{pmatrix}}_{\boldsymbol{A}_{\mathrm{c}i}} \underbrace{\begin{pmatrix} \varphi_i(t) \\ \dot{\varphi}_i(t) \end{pmatrix}}_{\boldsymbol{x}_{\mathrm{c}i}(t)} + \underbrace{\begin{pmatrix} 0 \\ -\frac{1}{M_i l_i} \end{pmatrix}}_{\boldsymbol{b}_{\mathrm{c}i}} \underbrace{F_i(t-\tau_{\mathrm{IO}i})}_{u_{\mathrm{c}i}(t-\tau_{\mathrm{IO}i})} + \underbrace{\begin{pmatrix} 0 \\ -\frac{1}{M_i l_i} \end{pmatrix}}_{\boldsymbol{b}_{\mathrm{dc}i}} \underbrace{F_{\mathrm{d}i}(t)}_{d_i(t)} \tag{10.1}$$

where $F_{\mathrm{d}i}$ is a disturbance force, refer to e.g. [KS72, Section 1.2.3] for a derivation. The inverted pendulums have the same pendulum mass $m_i = 0.1\,\mathrm{kg}$ (concentrated at the tip) and cart mass $M_i = 0.1\,\mathrm{kg}$ but different pendulum length $l_1 = 0.136\,\mathrm{m}$, $l_2 = 0.242\,\mathrm{m}$ and $l_3 = 0.545\,\mathrm{m}$. Friction is neglected and $g = 9.81\,\mathrm{m/s^2}$ is the gravitational acceleration. The eigenvalues of the inverted pendulums are

$$\lambda_{11/12} = \pm 12, \quad \lambda_{21/22} = \pm 9, \quad \lambda_{31/32} = \pm 6. \tag{10.2}$$

10.2.3 Selection of Weighting Matrices and Sampling Period

The weighting matrices $\boldsymbol{Q}_{1ci}$ and Q_{2ci} of the continuous-time cost function (7.24) and the sampling periods h_i are selected by designing a continuous-time controller which minimizes the cost function (7.24) for $T \to \infty$, $\boldsymbol{Q}_{0ci} = \boldsymbol{0}$ and $\tau_{IOi} = 0$, yielding a linear-quadratic regulator (LQR).

After several design iterations the weighting matrices are selected as

$$\boldsymbol{Q}_{1ci} = \begin{pmatrix} 10000 & 0 \\ 0 & 1 \end{pmatrix}, \quad Q_{2ci} = 1, \tag{10.3}$$

leading to the closed-loop eigenvalues

$$\tilde{\lambda}_{11/12} = -71.42 \pm j47.47, \quad \tilde{\lambda}_{21/22} = -50.34 \pm j39.99, \quad \tilde{\lambda}_{31/32} = -31.93 \pm j28.56. \tag{10.4}$$

Various methods for selecting the sampling period have been proposed in the literature, see [ÅW90, Sections 9.2 and 11.5] and [FPW90, Chapter 10] for a survey. Most methods rely on the rise time or bandwidth of the continuous-time closed-loop system. Moreover, if the continuous-time controller is designed based on minimizing a continuous-time cost function, then the effect of the sampling period on the cost can be quantified via the discretized cost function.

Requiring 4 to 10 samples per rise of the continuous-time closed-loop system as proposed in [ÅW90, Sections 9.2], appropriate sampling periods are

$$h_1 = 3\ldots 7\,\text{ms}, \quad h_2 = 4\ldots 9\,\text{ms}, \quad h_3 = 5\ldots 13\,\text{ms}, \tag{10.5}$$

yielding a cost degradation w.r.t. the cost for the continuous-time controller of

$$\text{P}_1: \; 0.4\ldots 2.1\,\%, \quad \text{P}_2: \; 0.4\ldots 2.0\,\%, \quad \text{P}_3: \; 0.3\ldots 1.9\,\%. \tag{10.6}$$

The sampling periods under the NECS model from Section 7.3 are determined by the task sequence $j(k)$ and the overall delays $\tau_i = 3.5\,\text{ms}$. Considering a 3-periodic task sequence $j(k) = (1, 2, 3)$, the sampling periods result as

$$h_i = 10.5\,\text{ms}, \tag{10.7}$$

yielding a cost degradation w.r.t. the cost for the continuous-time controller of

$$\text{P}_1: \; 4.6\,\%, \quad \text{P}_2: \; 2.7\,\%, \quad \text{P}_3: \; 1.2\,\%. \tag{10.8}$$

The cost degradations are clearly acceptable although the upper bounds given in (10.5) are slightly exceeded for P_1 and P_2. The pendulum angles $\varphi(t)$ and angular velocities $\dot{\varphi}(t)$ resulting for the continuous-time (CT) and discrete-time (DT) controller and the initial states $(\varphi_i(0) \;\; \dot{\varphi}_i(0))^T = \begin{pmatrix}0.1\,\text{rad} & 0\end{pmatrix}^T$ and disturbance forces $F_{\text{d}i}(t) = 0$ are shown in Figure 10.2, again affirming that the sampling periods (10.7) are acceptable. Note that the pendulum angles and angular velocities resulting for the DT controller are shown in Figure 10.2 only for the discretization instants (staircase graph). Note that furthermore the input-output delay $\tau_{\text{IO}i}$ which may further degrade the cost has deliberately been neglected in the preceding considerations.

10.2.4 Disturbance Model

The disturbance force $F_{\text{d}i}$ consists of rectangular impulses. The rectangular impulses arise at the discretization instants t_k. The impulse duration is equal to the discretization interval $t_{k+1} - t_k = \tau_i = 3.5\,\text{ms}$. The impulse amplitude is uniformly distributed on the interval $[-100\,\text{N}, 100\,\text{N}]$. The duration between impulses is uniformly distributed on the interval $[101.5\,\text{ms}, 500.5\,\text{ms}]$.

The disturbance force $F_{\text{d}i}$ affects the inverted pendulum just like the control force F_i since $\boldsymbol{b}_{\text{dc}i} = \boldsymbol{b}_{\text{c}i}$. Note that only the angular acceleration $\ddot{\varphi}_i$ is influenced directly. The disturbance may therefore also be interpreted as an impulsive angular acceleration.

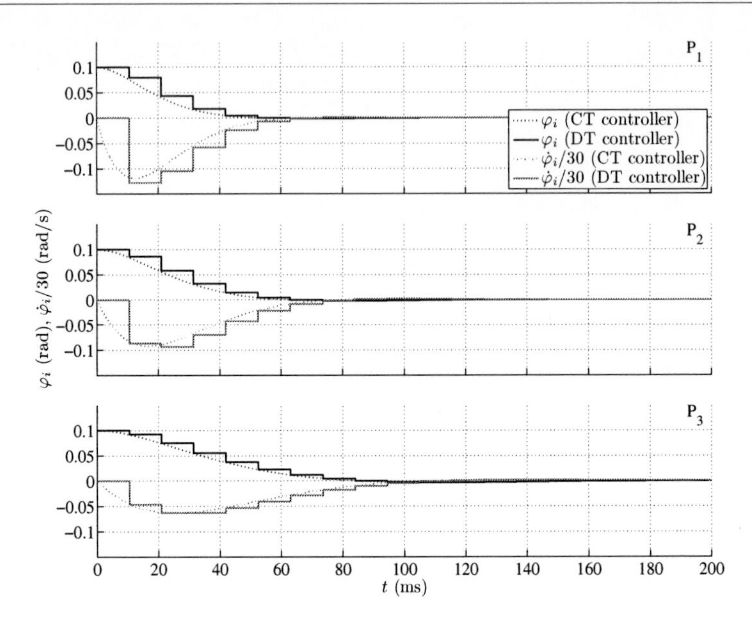

Figure 10.2: Pendulum angles φ_i and angular velocities $\dot{\varphi}_i$ for the continuous-time (CT) and discrete-time (DT) controller

10.3 Control and Scheduling Codesign

10.3.1 Conventional Control and Scheduling

Networked control of the three inverted pendulums is first addressed using a conventional control and scheduling strategy as commonly applied in practice. A 3-periodic task sequence $j(k) = (1, 2, 3)$ is considered which leads to the sampling periods $h_i = 10.5\,\text{ms}$. For these sampling periods, the continuous-time state equations (10.1) and cost functions (7.24) with the weighting matrices (10.3) are discretized using ZOH and neglecting the input-output delays $\tau_{\text{IO}i}$. Based on the resulting discrete-time state equations and cost functions discrete-time LQRs are designed.

A simulation of the NCS is performed utilizing the NECS model (7.21). Therefore, scheduling and communication which generally lead to time-varying sampling periods and time delays are regarded. Furthermore, force disturbances as specified in Section 10.2.4 and zero initial states are considered. The simulation time is chosen to $N_{\text{sim}} = 28572$ or equivalently $T_{\text{sim}} = 100.002\,\text{s}$. The cost is evaluated by reformulating the discretized overall cost function (7.30) to

$$J_{\text{sim}} = \sum_{k=0}^{N_{\text{sim}}} \left[\boldsymbol{x}^T(k)\boldsymbol{Q}_{1j(k)}\boldsymbol{x}(k) + 2\boldsymbol{x}^T(k)\boldsymbol{Q}_{12j(k)}\boldsymbol{u}(k) + \boldsymbol{u}^T(k)\boldsymbol{Q}_{2j(k)}\boldsymbol{u}(k) \right]. \tag{10.9}$$

The resulting cost is given by

$$J_{\text{sim}} = 13251. \tag{10.10}$$

10.3.2 Receding-Horizon Control and Scheduling

Networked control of the three inverted pendulums is second addressed using the RRHCS and N'-step RRHCS strategy. The prediction horizon is chosen to $N = 20$ or equivalently to $T = 70\,\text{ms}$ and thus covers the relevant dynamics as can be observed from Figure 10.2. The terminal weighting matrix $\boldsymbol{Q}_0$ is selected based on the PCS_{off} strategy using $p = 6$ and the expected cost (4.37).

The RRHCS and the N'-step RRHCS strategy are determined for different relaxation factors α using Algorithm 2.3, the N'-step RRHCS strategy also for different prediction subhorizons N'. For the N'-step RRHCS strategy furthermore a suboptimality analysis utilizing Algorithm 9.1 with an accuracy $\varepsilon = 0.001$ is performed. The resulting number of DRE solutions $|\mathbb{L}_0|$ before removing N'-step non-admissible task sequences, $|\mathbb{L}_0'|$ after removing N'-step non-admissible task sequences, their ratio $|\mathbb{L}_0'|/|\mathbb{L}_0|$ and $|\hat{\mathbb{L}}_0'|$ after suboptimality analysis which includes an additional relaxation, see Remark 9.6, are indicated in Table 10.1. Furthermore, the resulting suboptimality factor α' and the total suboptimality factor α_{tot} as well as the computation time T_{comp} for relaxed dynamic programming are given.

The number of DRE solutions $|\mathbb{L}_0|$, $|\mathbb{L}_0'|$ and $|\hat{\mathbb{L}}_0'|$ generally increases with decreasing α but remains acceptable even for small α. Remarkably, $|\hat{\mathbb{L}}_0'|$ is very small for all α. The ratio $|\mathbb{L}_0'|/|\mathbb{L}_0|$ generally increases with N', i.e. more task sequences are N'-step admissible as N' increases. As a consequence, the suboptimality induced by removing N'-step non-admissible task sequences which is quantified by the suboptimality factor α' generally decreases with increasing N', e.g. for $\alpha = 1.06$, affirming Remark 9.4.

Closed-loop stability of the NECS under the RRHCS strategy is examined based on Theorem 3.9 since only this a posteriori stability criterion is directly applicable for NECSs. Closed-loop stability of the NECS under the N'-step RRHCS strategy is analyzed based on Theorems 9.9 and 9.10. Fulfillment of Theorem 3.9 is indicated in Table 10.1 under SA, fulfillment of Theorem 9.9 or 9.10 under SA$'$.

Obviously, closed-loop stability under the RRHCS strategy can repeatedly be verified based on Theorem 3.9, even for large $|\mathbb{L}_0|$. On the other hand, closed-loop stability under the N'-step RRHCS strategy can always be verified based on Theorems 9.9 and 9.10, where notably always the more conservative Theorem 9.9 is already successful.

N'	α	$\lvert\mathbb{L}_0\rvert$	$\lvert\mathbb{L}'_0\rvert$	$\frac{\lvert\mathbb{L}'_0\rvert}{\lvert\mathbb{L}_0\rvert}$	$\lvert\hat{\mathbb{L}}'_0\rvert$	α'	α_{tot}	SA	SA'	J_{sim}	J'_{sim}	$\hat{J}'_{\text{sim}}$	T_{comp}
3	1.70	6	6	1.00	6	1.001	1.702	✓	✓	7454	8785	8785	10.99 s
3	1.50	12	6	0.50	6	1.040	1.560	✗	✓	7303	8785	8785	29.85 s
3	1.30	19	8	0.42	6	1.036	1.347	✗	✓	7276	8701	9265	1.11 min
3	1.18	41	13	0.32	6	1.060	1.250	✗	✓	7199	8925	9247	2.54 min
3	1.10	96	37	0.39	6	1.054	1.159	✗	✓	7196	8876	8975	9.29 min
3	1.09	107	36	0.34	6	1.054	1.149	✗	✓	7196	8952	9319	11.07 min
3	1.08	135	47	0.35	6	1.071	1.157	✓	✓	7195	8857	9379	13.51 min
3	1.07	157	51	0.32	6	1.083	1.159	✗	✓	7195	8891	9051	18.84 min
3	1.06	203	63	0.31	6	1.089	1.154	✗	✓	7195	8958	9011	32.75 min
3	1.05	266	81	0.30	6	1.069	1.123	✗	✓	7194	8797	9244	54.69 min
3	1.04	348	101	0.29	6	1.099	1.143	✓	✓	7194	9002	9337	1.72 h
4	1.70	6	6	1.00	6	1.001	1.702	✓	✓	7454	9400	9400	10.99 s
4	1.50	12	6	0.50	6	1.040	1.560	✗	✓	7303	9400	9400	29.85 s
4	1.30	19	12	0.63	7	1.032	1.342	✗	✓	7276	9558	8853	1.11 min
4	1.18	41	21	0.51	6	1.060	1.250	✗	✓	7199	9872	8926	2.54 min
4	1.10	96	60	0.63	6	1.054	1.159	✗	✓	7196	10148	8959	9.29 min
4	1.09	107	59	0.55	6	1.054	1.149	✗	✓	7196	9975	8908	11.07 min
4	1.08	135	74	0.55	8	1.042	1.125	✓	✓	7195	9833	8858	13.51 min
4	1.07	157	78	0.50	7	1.052	1.125	✗	✓	7195	9969	9045	18.84 min
4	1.06	203	104	0.51	10	1.036	1.098	✗	✓	7195	9729	9044	32.75 min
4	1.05	266	128	0.48	7	1.052	1.104	✗	✓	7194	10010	8952	54.69 min
4	1.04	348	162	0.47	8	1.050	1.092	✓	✓	7194	9802	8967	1.72 h
5	1.70	6	6	1.00	6	1.001	1.702	✓	✓	7454	9833	9833	10.99 s
5	1.50	12	6	0.50	6	1.040	1.560	✗	✓	7303	9833	9833	29.85 s
5	1.30	19	15	0.79	7	1.032	1.342	✗	✓	7276	10230	9552	1.11 min
5	1.18	41	27	0.66	6	1.060	1.250	✗	✓	7199	11194	9613	2.54 min
5	1.10	96	75	0.78	9	1.032	1.135	✗	✓	7196	10848	9661	9.29 min
5	1.09	107	76	0.71	6	1.054	1.149	✗	✓	7196	11318	9660	11.07 min
5	1.08	135	95	0.70	9	1.036	1.119	✓	✓	7195	11049	9672	13.51 min
5	1.07	157	103	0.66	7	1.052	1.125	✗	✓	7195	11268	9505	18.84 min
5	1.06	203	135	0.67	11	1.032	1.094	✗	✓	7195	10825	9777	32.75 min
5	1.05	266	175	0.66	7	1.052	1.104	✗	✓	7194	11121	9332	54.69 min
5	1.04	348	220	0.63	8	1.050	1.092	✓	✓	7194	11169	9607	1.72 h

Table 10.1: Results for the RRHCS and N'-step RRHCS strategy

Simulations of the NCS are performed as specified in Section 10.3.1. The costs J_{sim} resulting under the RRHCS strategy, J'_{sim} resulting under an N'-step RRHCS strategy considering all N'-step admissible task sequences and $\hat{J}'_{\text{sim}}$ resulting under an N'-step RRHCS strategy considering only the N'-step admissible task sequences remaining after the suboptimality analysis are given in Table 10.1.

The cost J_{sim} decreases with α. The decrease is marginal for $\alpha \leq 1.18$, however, the increase of $|\mathbb{L}_0|$ is significant for $\alpha \leq 1.18$. Hence, $\alpha = 1.18$ may be a reasonable choice. The costs J'_{sim} and $\hat{J}'_{\text{sim}}$ increase with N' which is attributed to the reduced adaptivity under disturbances since more steps of the control and task sequence have to be applied between reoptimizations for larger N'. Obviously, the loss in adaptivity overwhelms the gain in suboptimality pointed out above for increasing N'. Note that $J'_{\text{sim}} \leq \hat{J}'_{\text{sim}}$

for $N' = 3$ while $J'_{\text{sim}} \geq \hat{J}'_{\text{sim}}$ for $N' \in \{4, 5\}$, i.e. there is no evident advantage of considering all N'-step admissible task sequences. Furthermore, there is no obvious connection between J'_{sim}, $\hat{J}'_{\text{sim}}$ and α.

The task sequences j^* resulting under the RRHCS strategy for $\alpha = 1.04$ together with the disturbance forces $F_{\text{d}i}$ and pendulum angles φ_i for $t \in [5.3\,\text{s}, 6.1\,\text{s}]$ are shown in Figure 10.3. Evidently, a control task T_i is executed more frequently when the related pendulum P_i is subject to a disturbance impulse:

- For $t \in [5.3445\,\text{s}, 5.4145\,\text{s}]$, only P_2 is subject to a disturbance impulse. Consequently, only T_2 is executed.
- For $t \in [6.0025\,\text{s}, 6.1\,\text{s}]$, both P_2 and P_3 are subject to disturbance impulses. Consequently, T_2 and T_3 are executed alternately. Furthermore, T_1 is executed intermittently for completely compensating the disturbance impulse at $t = 5.9395\,\text{s}$.
- For $t \in [5.6\,\text{s}, 5.6805\,\text{s}]$, no pendulum subject to a disturbance impulse. Consequently, the task sequence j^* is essentially arbitrary.

These results impressively illustrate the adaptivity of the RRHCS strategy under disturbances. Similar results are obtained for all online scheduling strategies.

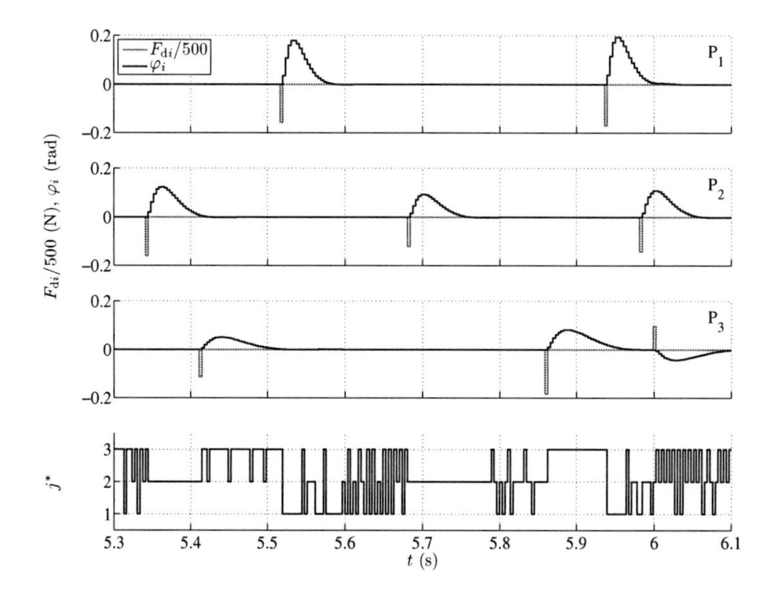

Figure 10.3: Task sequence j^*, disturbance forces $F_{\text{d}i}$ and pendulum angles φ_i

10.3.3 Periodic Control and Scheduling

Networked control of the three inverted pendulums is third addressed using the $\mathrm{PCS}_{\mathrm{off}}$, $\mathrm{PCS}_{\mathrm{on}}$, $\mathrm{RPCS}_{\mathrm{on}}$, $\mathrm{HPCS}_{\mathrm{on}}$ and OPP strategy. For all strategies simulations of the NCS are performed and the cost J_{sim} is computed as specified in Section 10.3.1. Since the study parallels the study in Examples 4.1, 4.2, 4.4, 4.6, 4.7 and 4.8, only the major results and differences are discussed in detail.

$\mathrm{PCS}_{\mathrm{off}}$ Strategy

The $\mathrm{PCS}_{\mathrm{off}}$ strategy is determined for different periods $p \in \{3, \ldots, 12\}$ utilizing Algorithm 4.1 with the expected cost $J^*_{p,\mathrm{exp}}$ according to (4.37). The number of p-periodic switching sequences $|\mathbb{J}_p|$, the number of admissible p-periodic switching sequences $|\mathbb{J}_{p,\mathrm{adm}}|$, their ratio $|\mathbb{J}_{p,\mathrm{adm}}|/|\mathbb{J}_p|$, the optimal admissible p-periodic switching sequence $j^*(k)$, the expected cost $J^*_{p,\mathrm{exp}}$ and the cost J_{sim} obtained for $j^*(k)$ and the computation time T_{comp} resulting for Algorithm 4.1 are indicated in Table 10.2.

p	$\lvert\mathbb{J}_p\rvert$	$\lvert\mathbb{J}_{p,\mathrm{adm}}\rvert$	$\frac{\lvert\mathbb{J}_{p,\mathrm{adm}}\rvert}{\lvert\mathbb{J}_p\rvert}$	$j^*(k)$	$J^*_{p,\mathrm{exp}}$	J_{sim}	T_{comp}
3	27	6	0.22	$(1,2,3)$	1028.41	10800	0.06 s
4	81	36	0.44	$(2,1,3,1)$	1032.31	10308	0.11 s
5	243	150	0.62	$(3,1,2,1,2)$	1030.63	10494	0.38 s
6	729	540	0.74	$(1,2,3,1,2,3)$	1028.41	10800	1.33 s
7	2187	1806	0.83	$(1,3,2,1,2,3,2)$	1028.66	11804	4.80 s
8	6561	5796	0.88	$(1,2,3,1,2,3,2,3)$	1028.01	12306	17.63 s
9	19683	18150	0.92	$(1,2,3,1,2,3,2,3,3)$	1028.11	14513	1.33 min
10	59049	55980	0.95	$(1,2,3,1,2,3,2,3,2,3)$	1027.84	16087	5.82 min
11	177147	171006	0.97	$(1,2,3,1,2,3,2,3,2,3,3)$	1027.72	18699	37.30 min
12	531441	519156	0.98	$(1,2,3,1,2,3,2,3,2,3,3,3)$	1027.62	20194	5.03 h

Table 10.2: Results for the $\mathrm{PCS}_{\mathrm{off}}$ strategy

The expected cost $J^*_{p,\mathrm{exp}}$ principally decreases with increasing period p but may also increase, e.g. between $p = 3$ and $p = 4$. Note that for $p \in \{3, 6\}$ the same $j^*(k)$ and therefore also the same $J^*_{p,\mathrm{exp}}$ is obtained. Essentially, already $p = 3$ may be a reasonable choice. The ratio $|\mathbb{J}_{p,\mathrm{adm}}|/|\mathbb{J}_p|$ increases with p which is generally the case for NECSs.

The optimal admissible p-periodic switching sequences $j^*(k)$ reflect the dynamics of the inverted pendulums. Pendulum P_1 having the fastest dynamics is controlled earlier, e.g. for $p = 3$, or more frequently, e.g. for $p = 4$, then pendulum P_2 having medium dynamics and then pendulum P_3 having the slowest dynamics.

The costs J_{sim} are similar for $p < 7$ but increase rapidly for $p \geq 7$. This is attributed to the fact that the $\mathrm{PCS}_{\mathrm{off}}$ strategy is optimal w.r.t. a given initial state according to (4.35), a random initial state according to (4.37) (considered here) or the worst-case initial state according to (4.38), but not w.r.t. persistent disturbances. Obviously, a kind of "overfitting" for the considered initial state is obtained for large p. This is also endorsed by the inhomogeneity of $j^*(k)$ for large p. Therefore, small p may be preferable under persistent disturbances.

$\mathbf{PCS_{on}}$ Strategy

The PCS_{on} strategy is determined for the period $p = 6$. The resulting cost is given by

$$J_{\text{sim}} = 7195 \tag{10.11}$$

and thus equal to the cost resulting under the RRHCS strategy for large α, cf. Table 10.1 but significantly smaller than the cost resulting under the PCS_{off} strategy, cf. Table 10.2. The online complexity may however be prohibitive since the region membership test comprises $|\mathbb{J}_{p,\text{adm}}| = 540$ regions, cf. Table 10.2.

$\mathbf{RPCS_{on}}$ Strategy

The $RPCS_{on}$ strategy is determined for the period $p = 6$ and different relaxation factors α utilizing Algorithms 4.2, 4.5 and 4.6. The number of admissible p-periodic switching sequences $|\hat{\mathbb{J}}_{p,\text{adm}}|$ before and after stability enforcement (SE) and the number of cyclically shifted admissible p-periodic switching sequences $v_{\text{s}}(k)$ that turned out to be already contained in $\hat{\mathbb{J}}_{p,\text{adm}}$ during SE and that turned out to be not contained in $\hat{\mathbb{J}}_{p,\text{adm}}$ during SE but had not to be added are indicated in Table 10.3. Moreover, the cost J_{sim} and overall computation time T_{comp} of the Algorithms are shown.

α	$\lvert\hat{\mathbb{J}}_{p,\text{adm}}\rvert$ after SE	$\lvert\hat{\mathbb{J}}_{p,\text{adm}}\rvert$ before SE	No. $v_{\text{s}}(i)$ contained	No. $v_{\text{s}}(i)$ not added	J_{sim}	T_{comp}
3.0000	9	2	2	0	7343	1.87 s
1.5000	12	6	6	0	7342	2.14 s
1.2500	12	6	6	0	7342	2.12 s
1.1000	12	6	6	0	7342	2.12 s
1.0750	12	6	6	0	7342	2.11 s
1.0500	18	7	7	0	7260	3.28 s
1.0250	42	14	14	0	7207	9.44 s
1.0115	126	44	44	0	7211	27.80 s
1.0100	156	54	54	0	7211	36.29 s
1.0075	174	70	70	0	7199	47.68 s
1.0050	258	113	113	0	7198	1.59 min
1.0025	372	193	193	0	7195	43.36 min
1.0010	450	288	288	0	7195	5.66 min
1.0000	540	538	538	0	7195	8.70 min

Table 10.3: Results for the $RPCS_{on}$ strategy ($p = 6$)

The number of admissible p-periodic switching sequences $|\hat{\mathbb{J}}_{p,\text{adm}}|$ after SE can be reduced significantly in comparison to the number of admissible p-periodic switching sequences

$|\mathbb{J}_{p,\mathrm{adm}}| = 540$ resulting under the $\mathrm{PCS_{on}}$ strategy without overly increasing the cost J_{sim}. A reasonable compromise between online complexity (determined by $|\hat{\mathbb{J}}_{p,\mathrm{adm}}|$ after SE) and cost may be given for $\alpha = 1.025$.

The cost J_{sim} resulting under the $\mathrm{RPCS_{on}}$ strategy is often slightly larger than under the RRHCS strategy for a comparable online complexity, cf. Tables 10.3 and 10.1 e.g. for $|\mathbb{J}_{p,\mathrm{adm}}| \in \{12, 42, 156\}$ and $|\mathbb{L}_0| \in \{12, 41, 157\}$, but not always, e.g. for $|\mathbb{J}_{p,\mathrm{adm}}| = 18$ and $|\mathbb{L}_0| = 19$.

$\mathrm{HPCS_{on}}$ Strategy

The $\mathrm{HPCS_{on}}$ strategy is determined for the period $p = 6$ and different cardinalities $M' \in \{6, 12, 42\}$ using Algorithm 4.1 with the expected cost $J^*_{p,\mathrm{exp}}$ according to (4.37). Since the determination of the $\mathrm{HPCS_{on}}$ strategy involves randomness, hundred $\mathrm{HPCS_{on}}$ strategies for each $M' \in \{6, 12, 42\}$ have been determined. The minimum, mean and maximum of the cost J_{sim} over the hundred strategies and the mean computation time $\overline{T}_{\mathrm{comp}}$ for determining the strategies are listed in Table 10.4.

M'	J_{sim} (minimum)	J_{sim} (mean)	J_{sim} (maximum)	$\overline{T}_{\mathrm{comp}}$
6	7580	7773	8044	3.59 s
12	7289	7334	7466	3.79 s
42	7212	7223	7236	4.98 s

Table 10.4: Results for the $\mathrm{HPCS_{on}}$ strategy ($p = 6$, $J^*_{p,\mathrm{exp}}$)

OPP Strategy

The OPP strategy is determined for the period $p = 6$ and utilizing Algorithm 4.1 with the expected cost $J^*_{p,\mathrm{exp}}$ according to (4.37). The resulting cost is given by

$$J_{\mathrm{sim}} = 7458. \tag{10.12}$$

10.4 Conclusions

Networked control of three inverted pendulums based on the control and scheduling codesign strategies proposed in Chapters 8 and 9 has been studied in this section.

The main results are summarized in Table 10.5. The online complexity characterized by the number of regions, the cost $J_{\mathrm{sim,strat}}$, the cost relation $J_{\mathrm{sim,strat}}/J_{\mathrm{sim,min}}$ w.r.t. the minimal achieved cost $J_{\mathrm{sim,min}}$ and the overall computation time T_{comp} are indicated.

The conventional control and scheduling strategy commonly applied in practice leads to a very large cost which strongly indicates the need for control and scheduling codesign.

Strategy		Online Complexity	$J_{\text{sim,strat}}$	$\frac{J_{\text{sim,strat}}}{J_{\text{sim,min}}}$	T_{comp}
Conventional		negligible	13251	1.8420	negligible
PCS_{off}	$(p=6,\ J^*_{p,\text{exp}})$	negligible	10800	1.5013	1.33 s
RRHCS	$(N=20,\ \alpha=1.04)$	$\lvert\mathbb{L}_0\rvert=348$	7194	1.0000	1.72 h
PCS_{on}	$(p=6)$	$\lvert\hat{\mathbb{J}}_{p,\text{adm}}\rvert=540$	7195	1.0001	1.33 s
RRHCS	$(N=20,\ \alpha=1.7)$	$\lvert\mathbb{L}_0\rvert=6$	7454	1.0361	10.99 s
N'-step RRHCS	$(N=20,\ N'=3,\ \alpha=1.7)$	$\lvert\mathbb{L}'_0\rvert=\lvert\hat{\mathbb{L}}'_0\rvert=6$	8785	1.2212	14.24 s
HPCS_{on}	$(p=6,\ J^*_{p,\text{exp}},$ minimum)	$M'=6$	7580	1.0537	4.92 s
HPCS_{on}	$(p=6,\ J^*_{p,\text{exp}},$ mean)	$M'=6$	7773	1.0805	4.92 s
OPP	$(p=6,\ J^*_{p,\text{exp}})$	$p=6$	7458	1.0367	1.33 s
RRHCS	$(N=20,\ \alpha=1.5)$	$\lvert\mathbb{L}_0\rvert=12$	7303	1.0152	1.11 min
RPCS_{on}	$(p=6,\ \alpha=1.075)$	$\lvert\hat{\mathbb{J}}_{p,\text{adm}}\rvert=12$	7342	1.0206	3.44 s
HPCS_{on}	$(p=6,\ J^*_{p,\text{exp}},$ minimum)	$M'=12$	7289	1.0132	5.12 s
HPCS_{on}	$(p=6,\ J^*_{p,\text{exp}},$ mean)	$M'=12$	7334	1.0195	5.12 s
RRHCS	$(N=20,\ \alpha=1.18)$	$\lvert\mathbb{L}_0\rvert=41$	7199	1.0007	2.54 min
RPCS_{on}	$(p=6,\ \alpha=1.025)$	$\lvert\hat{\mathbb{J}}_{p,\text{adm}}\rvert=42$	7207	1.0018	10.77 s
HPCS_{on}	$(p=6,\ J^*_{p,\text{exp}},$ minimum)	$M'=42$	7212	1.0025	6.31 s
HPCS_{on}	$(p=6,\ J^*_{p,\text{exp}},$ mean)	$M'=42$	7223	1.0040	6.31 s

Table 10.5: Comparison of the Control and Scheduling Codesign Strategies

The PCS_{off} strategy leads to a considerably smaller cost than the conventional control and scheduling strategy which is attributed to respecting the inverted pendulum dynamics in the task sequence on the one hand and regarding the transmission delays on the other hand. The online complexity is negligible.

The RRHCS strategy for $\alpha = 1.04$ and the PCS_{on} strategy lead to the smallest cost. The online complexity is, however, considerable.

The RRHCS, RPCS_{on}, HPCS_{on} and OPP strategy allow for a compromise between the online complexity and the cost. The obtained costs are grouped in Table 10.5 into three comparable online complexities.

The RRHCS strategy generally leads to the smallest cost. The RPCS_{on} strategy yields slightly larger costs, however, closed-loop stability is always guaranteed inherently. The costs resulting under the HPCS_{on} strategy strongly depends on the inherent randomness. However, very small costs can be achieved. The OPP strategy leads to a very small cost for $p = 6$ but can not be determined for a larger number of regions within reasonable computation time to further reduce the cost. The N'-step RRHCS strategy leads to substantial cost due to the poor adaptivity under persistent disturbances.

The overall computation times T_{comp} for the RHCS strategies are considerably larger than for the PCS strategies. This may, however, revert for NECSs with a larger number of plants. Furthermore, the PCS strategies partly rely on optimized algorithms, e.g. for solving algebraic Riccati equations, while the RHCS strategies rely on non-optimized algorithms. A more efficient implementation of the algorithms for the RHCS strategies may considerably speed up the computation.

The disturbance force F_{d1} and the pendulum angle φ_1 of pendulum P_1 resulting under the conventional control and scheduling strategy, the PCS_{off} strategy and the RRHCS strategy for $\alpha \in \{1.7, 1.3, 1.18, 1.04\}$ are shown in Figure 10.4 for $t \in [5.5\,s, 5.8\,s]$.

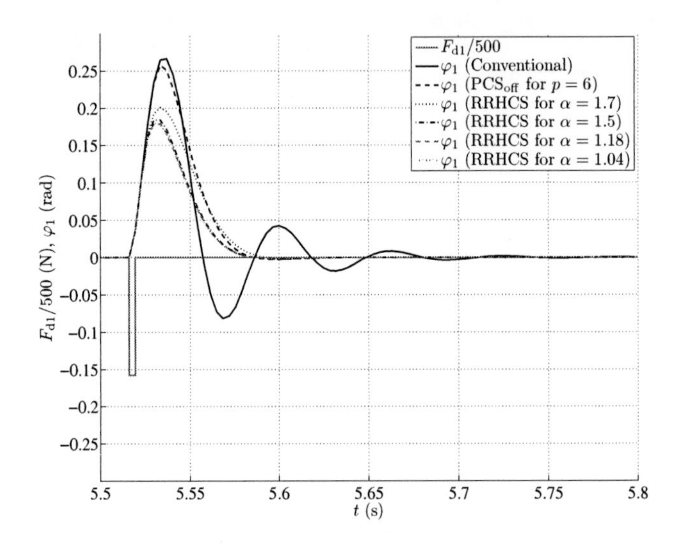

Figure 10.4: Disturbance force F_{d1} and pendulum angle φ_1 of pendulum P_1

The curves qualitatively reflect the costs: The performance under conventional control and scheduling is very poor. The performance under the PCS_{off} strategy is considerably better, particularly w.r.t. damping. The performance under the RRHCS strategy is very good for $\alpha = 1.7$, i.e. for only 6 regions, and excellent for $\alpha = 1.5$, i.e. for 12 regions, while the improvements for smaller $\alpha \in \{1.18, 1.04\}$ are only minor.

In summary, all control and scheduling codesign strategies are applicable for networked control of the three inverted pendulums. The performance ranges from acceptable to excellent, the online complexity from large to negligible. The RRHCS strategy and the $RPCS_{on}$ strategy allow to systematically balance between performance and online complexity and may therefore be the premier choice for control and scheduling codesign.

11 Conclusions and Future Work

11.1 Conclusions

A general framework for control and scheduling codesign of NECSs has been presented in Chapters 7 to 10. An NECS architecture consisting of multiple plants controlled over a single network by a single processor has been introduced. Such an NECS architecture is generic for NECSs with shared computation and communication resources and also frequently encountered in practice. For this NECS architecture a computation and communication model has been established which integrates computation, communication and control. This model is well suited for ECSs and NECSs where only controllers and actuators are connected over a network but may induce considerable scheduling overhead in NECSs where also sensors and controllers are connected over a network. Ideas to overcome this limitation are discussed in Section 11.2. Furthermore, an NECS model and cost function complementing the computation and communication model have been elaborated. Computation and communication times are represented in this NECS model and cost function in a very flexible way, allowing to describe the timing characteristics of different control algorithms and for different packet sizes precisely. The NECS model is represented by a block-diagonal discrete-time switched linear system with the switching index as interface to scheduling and medium access control and the control vector as interface to control. Strategies for control and scheduling codesign have been deduced from strategies for optimal control of switched systems proposed in Chapters 2 to 4. These strategies allow balancing between complexity, in particular scheduling overhead, and performance in a very systematic way. The structural properties of the NECS model have been thoroughly studied. Particularly, it has been revealed that the subsystems of the NECS model are not stabilizable. This property is regularly encountered in NECSs with shared computation and communication resources since plants are temporarily unattended. N'-step receding-horizon control and scheduling (N'-step RHCS) has been proposed to overcome this limitation, uniquely combining ideas from receding-horizon control and lifting. N'-step RHCS may also be applicable for other switched systems with non-stabilizable subsystems and is therefore relevant under a wider scope. Performance degradations are, however, inevitable under N'-step RHCS due to reduced reoptimizations and therefore reduced adaptivity under disturbances. Furthermore, suboptimality analysis has been addressed. A method for analyzing the suboptimality induced by removing switching sequences has been devised based on relaxation. This method is always applicable if switching sequences are removed which can inspire further approaches for complexity reduction with suboptimality bound. The effectiveness of control and scheduling codesign has been illustrated in a case study. Performance could be significantly improved in comparison to conventional control and scheduling as commonly used in practice. It could also be demonstrated that online scheduling, even with small scheduling overhead, usually provides better performance than offline scheduling.

11.2 Future Work

Robust Control and Scheduling Codesign

The computation and transmission times in NECSs are often uncertain as discussed in Section 6.1. Furthermore, packet loss may occur. While these computation and communication imperfections have been widely studied in terms of implementation-aware control, robust control and scheduling codesign has been rarely addressed. Future work may therefore focus on extending the models presented in Chapter 7 and the control and scheduling codesign strategies presented in Chapters 8 and 9 for NECSs with uncertain transmission and computation times as well as packet loss.

Robust control and scheduling codesign of NECSs with uncertain but interval-bounded time-varying computation and transmission times has been recently studied in [AGL11]. The NECS is modeled as a block-diagonal discrete-time switched linear system with polytopic and norm-bounded uncertainty, combining concepts from Chapter 7 to represent scheduling and from [IGL08, IGL10a, IGL10b] to represent uncertain time-varying computation and transmission times. Robust control and scheduling codesign is performed by extending the $\mathrm{PCS_{on}}$ strategy proposed in Section 4.2 to a Robust $\mathrm{PCS_{on}}$ strategy. The periodic control subproblem is solved based on a periodic parameter-dependent Lyapunov function, yielding an LMI optimization problem. The periodic scheduling subproblem is tackled based on exhaustive search. This concept may be extended for NECSs with uncertain-time varying sampling periods stemming e.g. from packet loss and preemptions as outlined in Section 6.1. Moreover, the RHCS, RRHCS and N'-step RHCS strategy may be revised for robust control and scheduling codesign.

Scheduling of the Sensor-Controller Link

Scheduling in the computation and communication model presented in Section 7.2 is based on the state vectors of all plants. Transmitting all state vectors, however, induces considerable scheduling overhead as outlined in Remark 7.2. Strategies for reducing the scheduling overhead are clearly required. Ideas include

- *Estimation*

  Estimation can be used to reconstruct the state vectors for scheduling. Only some state vectors must then be transmitted at each discretization instant. Persistent transmissions of all state vectors must, however, be assured to guarantee stability and performance, particularly under disturbances. To this end, centralized time-triggered hybrid scheduling, i.e. offline scheduling of the sensor-controller link and online scheduling of the controller-actuator link, may be considered. Moreover, event-triggered and stochastic [GCHM06] scheduling of the sensor-controller link may be investigated. Under such strategies only some new state vectors are available at each discretization instant. The estimator must thus run partly open loop. Novel methods for designing such estimators are required. Similar problems are encountered in Kalman filtering with intermittent observations [SSF+04], distributed Kalman filtering [Olf07, etc.], estimation over lossy networks [SSF+07, etc.] and sensor scheduling [AR05, GCHM06, VZA+10, etc.]. These concepts may provide a good basis for designing such estimators. Particularly the concepts presented in [VZA+10, AR05] which are based on dynamic programming with pruning and relaxed dynamic programming may be readily combined with RHCS.

- *Distributed Optimization*

  Distributed optimization can be considered for decentralized scheduling based on local information. Transmitting all state vectors can be avoided thereby. Various methods for distributed optimization have been proposed, see [YJ10] for a recent survey. Particularly methods for distributed receding-horizon control [BB10] may be inspected to extend the methods for control and scheduling codesign presented in Chapters 7 to 9 towards decentralized scheduling. Similar issues are also encountered in consensus problems, see [OFM07] for an overview.

NECS Architecture

The NECS architecture presented in Section 7.1 can be extended in several directions:

- NECSs with MIMO systems
  (Control and measurement data is split into several packets)
- NECSs with multiple networks and multiple processors
- NECSs with time-varying architecture
  (Number of processors and networks is time-varying)
- NECSs with energy constraints
  (Energy consumption for computation and transmission must be regarded)
- NECSs with security constraints
  (Networks and processors may be temporarily unavailable, data may be corrupted)

Such architectures become more and more important as NECS evolve to cyber-physical systems where ubiquity, connectivity, energy consumption and security are fundamental.

The models and control and scheduling codesign strategies must be extended for such architectures. While extensions for MIMO systems, multiple networks and multiple processors are quite straightforward, extensions for time-varying architectures, energy constraints and security constraints require fundamental research.

Part III

Appendix

A Supplementary Material

A.1 Expected Value of a Quadratic Form

Lemma A.1 *Let* $\boldsymbol{x} \in \mathbb{R}^{\mathrm{n}}$ *be a Gaussian random variable with expected value* $\mathrm{E}(\boldsymbol{x}) = 0$ *and covariance matrix* $\mathrm{E}(\boldsymbol{x}\boldsymbol{x}^T) = \boldsymbol{I}$.

The expected value of the quadratic form $\boldsymbol{x}^T\boldsymbol{P}\boldsymbol{x}$ *with* $\boldsymbol{P} \in \mathbb{R}^{\mathrm{n}\times\mathrm{n}}$ *is then given by*

$$\mathrm{E}(\boldsymbol{x}^T\boldsymbol{P}\boldsymbol{x}) = \mathrm{tr}(\boldsymbol{P}). \tag{A.1}$$

PROOF. It holds that

$$\mathrm{E}(\boldsymbol{x}^T\boldsymbol{P}\boldsymbol{x}) = \mathrm{E}\left(\sum_{i=1}^{\mathrm{n}}\sum_{j=1}^{\mathrm{n}} p_{ij}x_ix_j\right) = \sum_{i=1}^{\mathrm{n}}\sum_{j=1}^{\mathrm{n}} p_{ij}\mathrm{E}(x_ix_j) = \sum_{i=1}^{\mathrm{n}} p_{ii} = \mathrm{tr}(\boldsymbol{P}) \tag{A.2}$$

where the second equality follows from

$$\mathrm{E}\left(\sum_{i=1}^{\mathrm{n}} x_i\right) = \sum_{i=1}^{\mathrm{n}} \mathrm{E}(x_i), \quad \mathrm{E}(ax_i) = a\mathrm{E}(x_i) \tag{A.3}$$

and the third equality from

$$\mathrm{E}(\boldsymbol{x}\boldsymbol{x}^T) = \boldsymbol{I} \quad \Leftrightarrow \quad \mathrm{E}(x_ix_j) = \begin{cases} 0, & \text{for } i \neq j \\ 1, & \text{for } i = j. \end{cases} \tag{A.4}$$

□

Remark A.1. A more general treatment is given in [ÅW90, p. 338].

A.2 Minimum and Maximum Value of a Quadratic Form

Lemma A.2 *The minimum and maximum value of the quadratic form* $\boldsymbol{x}^T\boldsymbol{P}\boldsymbol{x}$ *on the unit hypersphere* $\mathcal{S} = \{\boldsymbol{x} \in \mathbb{R}^{\mathrm{n}} | \, \|\boldsymbol{x}\|_2 = 1\}$ *for* $\boldsymbol{P} \in \mathbb{R}^{\mathrm{n}\times\mathrm{n}}$ *symmetric is given by*

$$\min_{\boldsymbol{x}\in\mathcal{S}} \boldsymbol{x}^T\boldsymbol{P}\boldsymbol{x} = \lambda_{\mathsf{min}}(\boldsymbol{P}) \tag{A.5a}$$

$$\max_{\boldsymbol{x}\in\mathcal{S}} \boldsymbol{x}^T\boldsymbol{P}\boldsymbol{x} = \lambda_{\mathsf{max}}(\boldsymbol{P}) \tag{A.5b}$$

where $\lambda_{\min}(\cdot)$ *and* $\lambda_{\max}(\cdot)$ *denote the smallest and largest eigenvalue respectively. This leads to the Rayleigh-Ritz inequality*

$$\lambda_{\min}(\boldsymbol{P})\|\boldsymbol{x}\|_2^2 \leq \boldsymbol{x}^T\boldsymbol{P}\boldsymbol{x} \leq \lambda_{\max}(\boldsymbol{P})\|\boldsymbol{x}\|_2^2. \tag{A.6}$$

PROOF. See e.g. [Mey00, Example 7.5.1] and [HJ85, Theorem 4.2.2] □

A.3 Regions for Second-Order Switched Systems

Theorem A.3 *For second-order switched systems the regions* $\mathcal{X}_m$ *with* $m \in \mathbb{L}_0$ *are piecewise convex, i.e. they are composed of convex subregions. The boundaries of the regions are given by lines through the origin.*

PROOF. Recall that the boundary between two regions $\mathcal{X}_m$ and $\mathcal{X}_n$ with $m, n \in \mathbb{L}_0$ is given by (3.7). Inspect

$$\boldsymbol{x}^T\boldsymbol{P}_m\boldsymbol{x} = \boldsymbol{x}^T\boldsymbol{P}_n\boldsymbol{x} \quad \Leftrightarrow \quad \boldsymbol{x}^T\underbrace{(\boldsymbol{P}_m - \boldsymbol{P}_n)}_{\Delta\boldsymbol{P}_{mn}}\boldsymbol{x} = 0 \tag{A.7}$$

where $\Delta\boldsymbol{P}_{mn}$ is symmetric and indefinite since otherwise $\boldsymbol{P}_m$ or $\boldsymbol{P}_n$ would never be optimal. Further consider

$$\boldsymbol{x}^T\Delta\boldsymbol{P}_{mn}\boldsymbol{x} = 0 \quad \Leftrightarrow \quad \underbrace{\boldsymbol{x}^T\boldsymbol{V}^T}_{\boldsymbol{z}^T}\boldsymbol{D}\underbrace{\boldsymbol{V}\boldsymbol{x}}_{\boldsymbol{z}} = 0 \tag{A.8}$$

where the orthogonal matrix $\boldsymbol{V}$ and the diagonal matrix $\boldsymbol{D}$ follow by eigendecomposition of the symmetric matrix $\Delta\boldsymbol{P}_{mn}$, see e.g. [GV96, Section 8.1]. Rewriting (A.8) for a second-order system yields

$$\begin{pmatrix} z_1 & z_2 \end{pmatrix}\begin{pmatrix} \lambda_1 & 0 \\ 0 & \lambda_2 \end{pmatrix}\begin{pmatrix} z_1 \\ z_2 \end{pmatrix} = 0 \quad \Leftrightarrow \quad \lambda_1 z_1^2 + \lambda_2 z_2^2 = 0 \quad \Leftrightarrow \quad z_2 = \underbrace{\pm\sqrt{-\frac{\lambda_1}{\lambda_2}}}_{\pm\eta\in\mathbb{R}}\, z_1. \tag{A.9}$$

Consequently, $\boldsymbol{z}$ is described by the lines

$$\boldsymbol{z} = \begin{pmatrix} z_1 \\ z_2 \end{pmatrix} = \zeta\begin{pmatrix} 1 \\ \pm\eta \end{pmatrix}, \ \zeta \in \mathbb{R} \tag{A.10}$$

and therewith $\boldsymbol{x}$ is described by the lines

$$\boldsymbol{x} = \boldsymbol{V}^{-1}\boldsymbol{z} = \zeta\boldsymbol{V}^{-1}\begin{pmatrix} 1 \\ \pm\eta \end{pmatrix}, \ \zeta \in \mathbb{R}. \tag{A.11}$$

The boundaries of the regions are thus lines through the origin, implying piecewise convexity of the regions. □

Remark A.2. The boundaries of the regions can be constructed for second-order switched systems based on the preceding proof. The lines given by (A.11), however, are only potential boundaries. The fulfillment of the inequality contained in (3.7) must be tested separately by substituting (A.11) for some arbitrary $\zeta \in \mathbb{R}$.

A.4 Discretization of the Cost Function

This appendix addresses the discretization of the continuous-time cost function (7.24).

The cost function must be discretized over the discretization interval $t_k \leq t < t_{k+1}$ using ZOH. When selecting the time horizon $T = \sum_{k=0}^{N-1} \tau_{j(k)}$, the cost function can be rewritten as

$$J_{Ni} = \boldsymbol{x}_{ci}^T(T)\boldsymbol{Q}_{0ci}\boldsymbol{x}_{ci}(T) + \sum_{k=0}^{N-1} \int_{t_k}^{t_{k+1}} \boldsymbol{x}_{ci}^T(t)\boldsymbol{Q}_{1ci}\boldsymbol{x}_{ci}(t) + \boldsymbol{u}_i^T(t - \tau_{IOi})\boldsymbol{Q}_{2ci}\boldsymbol{u}_i(t - \tau_{IOi})dt \tag{A.12}$$

with $t_0 = 0$, $t_N = T$ and $t_{k+1} - t_k = \tau_{j(k)}$ according to (7.7).

Remark A.3. The time horizon T depends on the task sequence $j(k)$ for a given N. This is not problematic if N is chosen such that the relevant dynamics of the plant is covered within the time horizon T since then variations of T relate to times t where the state vector $\boldsymbol{x}_{ci}(t)$ and the control vector $\boldsymbol{u}_i(t - \tau_{IOi})$ are small and therefore affect the cost function J_{Ni} only marginally.

For the discretization, the task T_i considered for discretization and the task $\mathrm{T}_{j(k)}$ executed within the discretization interval $t_k \leq t < t_{k+1}$ must be distinguished. To this end the concepts presented in Section 7.3 can be utilized.

Substituting the solution of the continuous-time state equation (7.13) and the generalized delay $\tau_{ij(k)}$ according to (7.11) into (A.12) leads to

$$\begin{aligned} J_{Ni} =& \boldsymbol{x}_{ci}^T(T)\boldsymbol{Q}_{0ci}\boldsymbol{x}_{ci}(T) + \\ & \sum_{k=0}^{N-1} \int_{t_k}^{t_{k+1}} \left[\mathrm{e}^{\boldsymbol{A}_{ci}(t-t_k)}\boldsymbol{x}_{ci}(t_k) + \int_{t_k}^{t} \mathrm{e}^{\boldsymbol{A}_{ci}(t-s)}\boldsymbol{B}_{ci}\boldsymbol{u}_i(s - \tau_{ij(k)})ds \right]^T \boldsymbol{Q}_{1ci} \\ & \qquad \left[\mathrm{e}^{\boldsymbol{A}_{ci}(t-t_k)}\boldsymbol{x}_{ci}(t_k) + \int_{t_k}^{t} \mathrm{e}^{\boldsymbol{A}_{ci}(t-s)}\boldsymbol{B}_{ci}\boldsymbol{u}_i(s - \tau_{ij(k)})ds \right] + \\ & \qquad \boldsymbol{u}_i^T(t - \tau_{ij(k)})\boldsymbol{Q}_{2ci}\boldsymbol{u}_i(t - \tau_{ij(k)})dt. \end{aligned} \tag{A.13}$$

Expanding (A.13), substituting the piecewise constant control vector according to (7.12), splitting the integrals and factoring out $\boldsymbol{x}_{ci}(t_k)$, $\boldsymbol{u}_i(t_{k-1})$ and $\boldsymbol{u}_i(t_k)$ yields

$$J_{Ni} = \boldsymbol{x}_{ci}^T(T)\boldsymbol{Q}_{0ci}\boldsymbol{x}_{ci}(T) + \sum_{k=0}^{N-1} [I_1(k) + I_2(k) + I_3(k) + I_4(k) + I_5(k)] \tag{A.14}$$

with

$$I_1(k) = \int_{t_k}^{t_{k+1}} \left[\mathrm{e}^{\boldsymbol{A}_{ci}(t-t_k)}\boldsymbol{x}_{ci}(t_k)\right]^T \boldsymbol{Q}_{1ci} \left[\mathrm{e}^{\boldsymbol{A}_{ci}(t-t_k)}\boldsymbol{x}_{ci}(t_k)\right] dt \tag{A.15a}$$

$$= \boldsymbol{x}_{ci}^T(t_k) \left[\int_0^{\tau_{j(k)}} \left(\mathrm{e}^{\boldsymbol{A}_{ci}t}\right)^T \boldsymbol{Q}_{1ci} \left(\mathrm{e}^{\boldsymbol{A}_{ci}t}\right) dt\right] \boldsymbol{x}_{ci}(t_k)$$

$$I_2(k) = \int_{t_k}^{t_{k+1}} \left[\mathrm{e}^{\boldsymbol{A}_{ci}(t-t_k)}\boldsymbol{x}_{ci}(t_k)\right]^T \boldsymbol{Q}_{1ci} \left[\int_{t_k}^{t} \mathrm{e}^{\boldsymbol{A}_{ci}(t-s)}\boldsymbol{B}_{ci}\boldsymbol{u}_i(s-\tau_{ij(k)})ds\right] dt \tag{A.15b}$$

$$= \int_{t_k}^{t_k+\tau_{ij(k)}} \left[\mathrm{e}^{\boldsymbol{A}_{ci}(t-t_k)}\boldsymbol{x}_{ci}(t_k)\right]^T \boldsymbol{Q}_{1ci} \left[\int_{t_k}^{t} \mathrm{e}^{\boldsymbol{A}_{ci}(t-s)}\boldsymbol{B}_{ci}\boldsymbol{u}_i(t_{k-1})ds\right] dt +$$

$$\int_{t_k+\tau_{ij(k)}}^{t_{k+1}} \left[\mathrm{e}^{\boldsymbol{A}_{ci}(t-t_k)}\boldsymbol{x}_{ci}(t_k)\right]^T \boldsymbol{Q}_{1ci}$$

$$\left[\int_{t_k}^{t_k+\tau_{ij(k)}} \mathrm{e}^{\boldsymbol{A}_{ci}(t-s)}\boldsymbol{B}_{ci}\boldsymbol{u}_i(t_{k-1})ds + \int_{t_k+\tau_{ij(k)}}^{t} \mathrm{e}^{\boldsymbol{A}_{ci}(t-s)}\boldsymbol{B}_{ci}\boldsymbol{u}_i(t_k)ds\right] dt$$

$$= \boldsymbol{x}_{ci}^T(t_k) \left[\int_0^{\tau_{ij(k)}} \left(\mathrm{e}^{\boldsymbol{A}_{ci}t}\right)^T \boldsymbol{Q}_{1ci} \left(\int_0^{t} \mathrm{e}^{\boldsymbol{A}_{ci}(t-s)}\boldsymbol{B}_{ci}ds\right) dt\right] \boldsymbol{u}_i(t_{k-1}) +$$

$$\boldsymbol{x}_{ci}^T(t_k) \left[\int_{\tau_{ij(k)}}^{\tau_{j(k)}} \left(\mathrm{e}^{\boldsymbol{A}_{ci}t}\right)^T \boldsymbol{Q}_{1ci} \left(\int_0^{\tau_{ij(k)}} \mathrm{e}^{\boldsymbol{A}_{ci}(t-s)}\boldsymbol{B}_{ci}ds\right) dt\right] \boldsymbol{u}_i(t_{k-1}) +$$

$$\boldsymbol{x}_{ci}^T(t_k) \left[\int_{\tau_{ij(k)}}^{\tau_{j(k)}} \left(\mathrm{e}^{\boldsymbol{A}_{ci}t}\right)^T \boldsymbol{Q}_{1ci} \left(\int_{\tau_{ij(k)}}^{t} \mathrm{e}^{\boldsymbol{A}_{ci}(t-s)}\boldsymbol{B}_{ci}ds\right) dt\right] \boldsymbol{u}_i(t_k)$$

$$I_3(k) = \int_{t_k}^{t_{k+1}} \left[\int_{t_k}^{t} \mathrm{e}^{\boldsymbol{A}_{ci}(t-s)}\boldsymbol{B}_{ci}\boldsymbol{u}_i(s-\tau_{ij(k)})ds\right]^T \boldsymbol{Q}_{1ci} \left[\mathrm{e}^{\boldsymbol{A}_{ci}(t-t_k)}\boldsymbol{x}_{ci}(t_k)\right] dt \tag{A.15c}$$

$$= \int_{t_k}^{t_k+\tau_{ij(k)}} \left[\int_{t_k}^{t} \mathrm{e}^{\boldsymbol{A}_{ci}(t-s)}\boldsymbol{B}_{ci}\boldsymbol{u}_i(t_{k-1})ds\right]^T \boldsymbol{Q}_{1ci} \left[\mathrm{e}^{\boldsymbol{A}_{ci}(t-t_k)}\boldsymbol{x}_{ci}(t_k)\right] dt +$$

$$\int_{t_k+\tau_{ij(k)}}^{t_{k+1}} \left[\int_{t_k}^{t_k+\tau_{ij(k)}} \mathrm{e}^{\boldsymbol{A}_{ci}(t-s)}\boldsymbol{B}_{ci}\boldsymbol{u}_i(t_{k-1})ds + \int_{t_k+\tau_{ij(k)}}^{t} \mathrm{e}^{\boldsymbol{A}_{ci}(t-s)}\boldsymbol{B}_{ci}\boldsymbol{u}_i(t_k)ds\right]^T \boldsymbol{Q}_{1ci}$$

$$\left[\mathrm{e}^{\boldsymbol{A}_{\mathrm{c}i}(t-t_k)}\boldsymbol{x}_{\mathrm{c}i}(t_k)\right] dt$$

$$=\boldsymbol{u}_i^T(t_{k-1})\left[\int_0^{\tau_{ij(k)}}\left(\int_0^t \mathrm{e}^{\boldsymbol{A}_{\mathrm{c}i}(t-s)}\boldsymbol{B}_{\mathrm{c}i}ds\right)^T \boldsymbol{Q}_{1\mathrm{c}i}\left(\mathrm{e}^{\boldsymbol{A}_{\mathrm{c}i}t}\right) dt\right]\boldsymbol{x}_{\mathrm{c}i}(t_k)+$$

$$\boldsymbol{u}_i^T(t_{k-1})\left[\int_{\tau_{ij(k)}}^{\tau_{j(k)}}\left(\int_0^{\tau_{ij(k)}} \mathrm{e}^{\boldsymbol{A}_{\mathrm{c}i}(t-s)}\boldsymbol{B}_{\mathrm{c}i}ds\right)^T \boldsymbol{Q}_{1\mathrm{c}i}\left(\mathrm{e}^{\boldsymbol{A}_{\mathrm{c}i}t}\right) dt\right]\boldsymbol{x}_{\mathrm{c}i}(t_k)+$$

$$\boldsymbol{u}_i^T(t_k)\left[\int_{\tau_{ij(k)}}^{\tau_{j(k)}}\left(\int_{\tau_{ij(k)}}^t \mathrm{e}^{\boldsymbol{A}_{\mathrm{c}i}(t-s)}\boldsymbol{B}_{\mathrm{c}i}ds\right)^T \boldsymbol{Q}_{1\mathrm{c}i}\left(\mathrm{e}^{\boldsymbol{A}_{\mathrm{c}i}t}\right) dt\right]\boldsymbol{x}_{\mathrm{c}i}(t_k)$$

$$I_4(k)=\int_{t_k}^{t_{k+1}}\left[\int_{t_k}^t \mathrm{e}^{\boldsymbol{A}_{\mathrm{c}i}(t-s)}\boldsymbol{B}_{\mathrm{c}i}\boldsymbol{u}_i(s-\tau_{ij(k)})ds\right]^T \boldsymbol{Q}_{1\mathrm{c}i}\left[\int_{t_k}^t \mathrm{e}^{\boldsymbol{A}_{\mathrm{c}i}(t-s)}\boldsymbol{B}_{\mathrm{c}i}\boldsymbol{u}_i(s-\tau_{ij(k)})ds\right] dt \tag{A.15d}$$

$$=\int_{t_k}^{t_k+\tau_{ij(k)}}\left[\int_{t_k}^t \mathrm{e}^{\boldsymbol{A}_{\mathrm{c}i}(t-s)}\boldsymbol{B}_{\mathrm{c}i}\boldsymbol{u}_i(t_{k-1})ds\right]^T \boldsymbol{Q}_{1\mathrm{c}i}\left[\int_{t_k}^t \mathrm{e}^{\boldsymbol{A}_{\mathrm{c}i}(t-s)}\boldsymbol{B}_{\mathrm{c}i}\boldsymbol{u}_i(t_{k-1})ds\right] dt+$$

$$\int_{t_k+\tau_{ij(k)}}^{t_{k+1}}\left[\int_{t_k}^{t_k+\tau_{ij(k)}} \mathrm{e}^{\boldsymbol{A}_{\mathrm{c}i}(t-s)}\boldsymbol{B}_{\mathrm{c}i}\boldsymbol{u}_i(t_{k-1})ds+\int_{t_k+\tau_{ij(k)}}^t \mathrm{e}^{\boldsymbol{A}_{\mathrm{c}i}(t-s)}\boldsymbol{B}_{\mathrm{c}i}\boldsymbol{u}_i(t_k)ds\right]^T \boldsymbol{Q}_{1\mathrm{c}i}$$

$$\left[\int_{t_k}^{t_k+\tau_{ij(k)}} \mathrm{e}^{\boldsymbol{A}_{\mathrm{c}i}(t-s)}\boldsymbol{B}_{\mathrm{c}i}\boldsymbol{u}_i(t_{k-1})ds+\int_{t_k+\tau_{ij(k)}}^t \mathrm{e}^{\boldsymbol{A}_{\mathrm{c}i}(t-s)}\boldsymbol{B}_{\mathrm{c}i}\boldsymbol{u}_i(t_k)ds\right] dt$$

$$=\boldsymbol{u}_i^T(t_{k-1})\left[\int_0^{\tau_{ij(k)}}\left(\int_0^t \mathrm{e}^{\boldsymbol{A}_{\mathrm{c}i}(t-s)}\boldsymbol{B}_{\mathrm{c}i}ds\right)^T \boldsymbol{Q}_{1\mathrm{c}i}\left(\int_0^t \mathrm{e}^{\boldsymbol{A}_{\mathrm{c}i}(t-s)}\boldsymbol{B}_{\mathrm{c}i}ds\right) dt\right]\boldsymbol{u}_i(t_{k-1})+$$

$$\boldsymbol{u}_i^T(t_{k-1})\left[\int_{\tau_{ij(k)}}^{\tau_{j(k)}}\left(\int_0^{\tau_{ij(k)}} \mathrm{e}^{\boldsymbol{A}_{\mathrm{c}i}(t-s)}\boldsymbol{B}_{\mathrm{c}i}ds\right)^T \boldsymbol{Q}_{1\mathrm{c}i}\left(\int_0^{\tau_{ij(k)}} \mathrm{e}^{\boldsymbol{A}_{\mathrm{c}i}(t-s)}\boldsymbol{B}_{\mathrm{c}i}ds\right) dt\right]\boldsymbol{u}_i(t_{k-1})+$$

$$\boldsymbol{u}_i^T(t_{k-1})\left[\int_{\tau_{ij(k)}}^{\tau_{j(k)}}\left(\int_0^{\tau_{ij(k)}} \mathrm{e}^{\boldsymbol{A}_{\mathrm{c}i}(t-s)}\boldsymbol{B}_{\mathrm{c}i}ds\right)^T \boldsymbol{Q}_{1\mathrm{c}i}\left(\int_{\tau_{ij(k)}}^t \mathrm{e}^{\boldsymbol{A}_{\mathrm{c}i}(t-s)}\boldsymbol{B}_{\mathrm{c}i}ds\right) dt\right]\boldsymbol{u}_i(t_k)+$$

$$\boldsymbol{u}_i^T(t_k)\left[\int_{\tau_{ij(k)}}^{\tau_{j(k)}}\left(\int_{\tau_{ij(k)}}^t \mathrm{e}^{\boldsymbol{A}_{\mathrm{c}i}(t-s)}\boldsymbol{B}_{\mathrm{c}i}ds\right)^T \boldsymbol{Q}_{1\mathrm{c}i}\left(\int_0^{\tau_{ij(k)}} \mathrm{e}^{\boldsymbol{A}_{\mathrm{c}i}(t-s)}\boldsymbol{B}_{\mathrm{c}i}ds\right) dt\right]\boldsymbol{u}_i(t_{k-1})+$$

$$\boldsymbol{u}_i^T(t_k)\left[\int\limits_{\tau_{ij(k)}}^{\tau_{j(k)}}\left(\int\limits_{\tau_{ij(k)}}^{t} \mathrm{e}^{\boldsymbol{A}_{\mathrm{c}i}(t-s)}\boldsymbol{B}_{\mathrm{c}i}ds\right)^T \boldsymbol{Q}_{1\mathrm{c}i}\left(\int\limits_{\tau_{ij(k)}}^{t} \mathrm{e}^{\boldsymbol{A}_{\mathrm{c}i}(t-s)}\boldsymbol{B}_{\mathrm{c}i}ds\right)dt\right]\boldsymbol{u}_i(t_k)$$

$$I_5(k) = \int\limits_{t_k}^{t_{k+1}} \boldsymbol{u}_i^T(t-\tau_{\tau_{ij(k)}})\boldsymbol{Q}_{2\mathrm{c}i}\boldsymbol{u}_i(t-\tau_{\tau_{ij(k)}})dt \tag{A.15e}$$

$$= \int\limits_{t_k}^{t_k+\tau_{ij(k)}} \boldsymbol{u}_i^T(t_{k-1})\boldsymbol{Q}_{2\mathrm{c}i}\boldsymbol{u}_i(t_{k-1})dt + \int\limits_{t_k+\tau_{ij(k)}}^{\tau_{j(k)}} \boldsymbol{u}_i^T(t_k)\boldsymbol{Q}_{2\mathrm{c}i}\boldsymbol{u}_i(t_k)dt$$

$$=\boldsymbol{u}_i^T(t_{k-1})\left[\int\limits_{0}^{\tau_{ij(k)}}\boldsymbol{Q}_{2\mathrm{c}i}dt\right]\boldsymbol{u}_i(t_{k-1}) + \boldsymbol{u}_i^T(t_k)\left[\int\limits_{\tau_{ij(k)}}^{\tau_{j(k)}}\boldsymbol{Q}_{2\mathrm{c}i}dt\right]\boldsymbol{u}_i(t_k).$$

Substituting the augmented state vector $\boldsymbol{x}_i(k)$ according to (7.19a), the matrices $\boldsymbol{\Phi}_{ij(k)}(t)$ and $\boldsymbol{\Gamma}_{ij(k)}(\underline{t},\overline{t})$ according to (7.27) and furthermore $k \in \mathbb{N}_0$ and N to abstractly represent t_k and $t_N = T$ finally yields the discrete-time cost function (7.25) with the weighting matrices (7.26).

Remark A.4. The discretization of the continuous-time cost function (7.24) has first been addressed in [San04, Paper B, Section 5.3 and Appendix D]. The derivation given there unfortunately contains some mistakes. These mistakes have been recognized by [Al-11] and corrected in the derivation given above.

Remark A.5. The computation of the weighting matrices (7.26) can be reformulated to the computation of matrix exponentials, see [FPW90, Section 9.4.4] for the general idea and [San04, Paper B, Appendix F] for details.

Remark A.6. The matrices $\boldsymbol{\Gamma}_{ij(k)}(\underline{t},\overline{t})$ in the weighting matrices (7.26) can be represented by

$$\boldsymbol{\Gamma}_{ij(k)}(0,t) = \tilde{\boldsymbol{\Gamma}}_{ij(k)}(t) \tag{A.16a}$$

$$\boldsymbol{\Gamma}_{ij(k)}(\tau_{ij(k)},t) = \tilde{\boldsymbol{\Gamma}}_{ij(k)}(t-\tau_{ij(k)}) \tag{A.16b}$$

$$\boldsymbol{\Gamma}_{ij(k)}(0,\tau_{ij(k)}) = \tilde{\boldsymbol{\Gamma}}_{ij(k)}(t) - \tilde{\boldsymbol{\Gamma}}_{ij(k)}(t-\tau_{ij(k)}) \tag{A.16c}$$

with

$$\tilde{\boldsymbol{\Gamma}}_{ij(k)}(t) = \int_0^t \mathrm{e}^{\boldsymbol{A}_{\mathrm{c}i}s}ds. \tag{A.17}$$

This representation is sometimes more convenient, e.g. for computing the weighting matrices (7.26) or for handling uncertain time-varying computation and transmission times as outlined in [IGL10a, AGL11].

Bibliography

[AB04] Panos J. Antsaklis and John Baillieul. Special issue on networked control systems. *IEEE Transactions on Automatic Control*, 49(9):1421–1423, 2004.

[AB07] Panos J. Antsaklis and John Baillieul. Special issue on technology of networked control systems. *Proceedings of the IEEE*, 95(1):5–8, 2007.

[AB09] Alessandro Alessio and Alberto Bemporad. A survey on explicit model predictive control. In Lalo Magni, Davide Martino Raimondo, and Frank Allgöwer, editors, *Nonlinear Model Predictive Control: Towards New Challenging Applications*, volume 384 of *Lecture Notes in Control and Information Sciences*, pages 99–108. Springer, 2009.

[ÅC05] Karl-Erik Årzén and Anton Cervin. Control and embedded computing: Survey of research directions. In *Proceedings of the 16th IFAC World Congress*, 2005.

[AGL11] Sanad Al-Areqi, Daniel Görges, and Steven Liu. Robust control and scheduling codesign for networked embedded control systems. In *Proceedings of the 50th IEEE Conference on Decision and Control and European Control Conference 2011*, pages 3154–3159, 2011.

[AK03] Panos J. Antsaklis and Xenofon D. Koutsoukos. Hybrid dynamical systems: Review and recent progress. In Tariq Samad and Gary Balas, editors, *Software-Enabled Control: Information Technology for Dynamical Systems*, chapter 14, pages 273–298. John Wiley & Sons, Hoboken, NJ, 2003.

[Al-11] Sanad Al-Areqi. Discretization of cost functions for systems with input delay. Personal communication, University of Kaiserslautern, Department of Electrical and Computer Engineering, Institute of Control Systems, Kaiserslautern, Germany, 2011.

[AN98] Panos J. Antsaklis and Anil Nerode. Hybrid control systems: An introductory discussion to the special issue. *IEEE Transactions on Automatic Control*, 43(4):457–460, 1998.

[Ant00] Panos J. Antsaklis. Special issue on hybrid systems: Theory and applications: A brief introduction to the theory and applications of hybrid systems. *Proceedings of the IEEE*, 88(7):879–887, 2000.

[AR05] Peter Alriksson and Anders Rantzer. Sub-optimal sensor scheduling with error bounds. In *Proceedings of the 16th IFAC World Congress*, pages 80–84, 2005.

[ÅW90] Karl-Johan Åström and Björn Wittenmark. *Computer-Controlled Systems: Theory and Design.* Prentice-Hall, Englewood Cliffs, NJ, 2nd edition, 1990.

[BA07] John Baillieul and Panos J. Antsaklis. Control and communication challenges in networked real-time systems. *Proceedings of the IEEE*, 95(1):9–28, 2007.

[BB85a] Sergio Bittanti and Paolo Bolzern. Discrete-time linear periodic systems: Gramian and modal criteria for reachability and controllability. *International Journal of Control*, 41(4):909–928, 1985.

[BB85b] Sergio Bittanti and Paolo Bolzern. Reachability and controllability of discrete-time linear periodic systems. *IEEE Transactions on Automatic Control*, 30(4):399–401, 1985.

[BB10] Alberto Bemporad and Davide Barcelli. Decentralized model predictive control. In Alberto Bemporad, Maurice Heemels, and Mikael Johansson, editors, *Networked Control Systems*, volume 406 of *Lecture Notes in Control and Information Sciences*, chapter 5, pages 149–178. Springer, Berlin, 2010.

[BBBM05] Francesco Borrelli, Mato Baotić, Alberto Bemporad, and Manfred Morari. Dynamic programming for constrained optimal control of discrete-time linear hybrid systems. *Automatica*, 41(10):1709–1721, 2005.

[BBD+00] Andrea Balluchi, Luca Benvenuti, Maria Domenica Di Benedetto, Claudio Pinello, and Alberto Luigi Sangiovanni-Vincentelli. Automotive engine control and hybrid systems: Challenges and opportunities. *Proceedings of the IEEE*, 88(7):888–912, 2000.

[BBE+05] Andrea Balluchi, Luca Benvenuti, Sebastian Engell, Tobias Geyer, Karl Henrik Johansson, Françoise Lamnabhi-Lagarrigue, John Lygeros, Manfred Morari, Georgios Papafotiou, Alberto Luigi Sangiovanni-Vincentelli, Fortunato Santucci, and Olaf Stursberg. Hybrid control of networked embedded systems. *European Journal of Control*, 11(4):478–508, 2005.

[BBM03] Alberto Bemporad, Francesco Borrelli, and Manfred Morari. Min-max control of constrained uncertain discrete-time linear systems. *IEEE Transactions on Automatic Control*, 48(9):1600–1606, 2003.

[BC00] Sergio Bittanti and Patrizio Colaneri. Invariant representations of discrete-time periodic systems. *Automatica*, 36(12):1777–1793, 2000.

[BC09] Sergio Bittanti and Patrizio Colaneri. *Periodic Systems: Filtering and Control.* Communications and Control Engineering. Springer, London, 2009.

[BCD88] Sergio Bittanti, Patrizio Colaneri, and Giuseppe De Nicolao. The difference periodic Riccati equation for the periodic prediction problem. *IEEE Transactions on Automatic Control*, 33(8):706–712, 1988.

[BCD91] Sergio Bittanti, Patrizio Colaneri, and Giuseppe De Nicolao. The periodic Riccati equation. In Sergio Bittanti, Alan J. Laub, and Jan C. Willems, editors, *The Riccati Equation*, Communications and Control Engineering Series, chapter 6, pages 127–162. Springer, Berlin, 1991.

[BÇH06] Mohamed El Mongi Ben Gaid, Arben Çela, and Yskandar Hamam. Optimal integrated control and scheduling of networked control systems with communication constraints: Application to a car suspension system. *IEEE Transactions on Control Systems Technology*, 14(4):776–787, 2006.

[BÇH09] Mohamed El Mongi Ben Gaid, Arben Çela, and Yskandar Hamam. Optimal real-time scheduling of control tasks with state feedback resource allocation. *IEEE Transactions on Control Systems Technology*, 17(2):309–326, 2009.

[BCM06] Mato Baotić, Frank J. Christophersen, and Manfred Morari. Constrained optimal control of hybrid systems with a linear performance index. *IEEE Transactions on Automatic Control*, 51(12):1903–1919, 2006.

[BD62] Richard E. Bellman and Stuart E. Dreyfus. *Applied Dynamic Programming.* Princeton University Press, Princeton, NJ, 1962.

[BE03] Mohamed Babaali and Magnus Egerstedt. Pathwise observability and controllability are decidable. In *Proceedings of the 42nd IEEE Conference on Decision and Control*, pages 5771–5776, 2003.

[BEFB94] Stephen Boyd, Laurent El Ghaoui, Eric Feron, and Venkataramanan Balakrishnan. *Linear Matrix Inequalities in System and Control Theory.* Society for Industrial and Applied Mathematics (SIAM), Philadelphia, PA, 1994.

[Bel57] Richard E. Bellman. *Dynamic Programming.* Princeton University Press, Princeton, NJ, 1957.

[Ber01] Dimitri P. Bertsekas. *Dynamic Programming and Optimal Control*, volume 2. Athena Scientific, Belmont, MA, 2nd edition, 2001.

[Ber05a] Dimitri P. Bertsekas. *Dynamic Programming and Optimal Control*, volume 1. Athena Scientific, Belmont, MA, 3rd edition, 2005.

[Ber05b] Dimitri P. Bertsekas. Dynamic programming and suboptimal control: A survey from ADP to MPC. *European Journal of Control*, 11(4-5):310–334, 2005.

[BF03] Alberto Bemporad and Carlo Filippi. Suboptimal explicit receding horizon control via approximate multiparametric quadratic programming. *Journal of Optimization Theory and Applications*, 117(1):9–38, 2003.

[BG11] Radhakisan Baheti and Helen Gill. Cyber-physical systems. In Tariq Samad and Anuradha Annaswamy, editors, *The Impact of Control Technology: Overview, Success Stories, and Research Challenges*, pages 161–166. IEEE Control Systems Society, 2011.

[BGLM05] Pratik Biswas, Pascal Grieder, Johan Löfberg, and Manfred Morari. A survey on stability analysis of discrete-time piecewise affine systems. In *Proceedings of 16th IFAC World Congress*, 2005.

[BGM05] A. Giovanni Beccuti, Tobias Geyer, and Manfred Morari. A hybrid system approach to power systems voltage control. In *Proceedings of the 44th IEEE Conference on Decision and Control and European Control Conference 2005*, pages 6774–6779, 2005.

[BGW90] Robert H. Bitmead, Michel Gevers, and Vincent Wertz. *Adaptive Optimal Control: The Thinking Man's GPC.* Prentice-Hall International Series in Systems and Control Engineering. Prentice Hall, New York, NY, 1990.

[BHJ10] Alberto Bemporad, Maurice Heemels, and Mikael Johansson, editors. *Networked Control Systems*, volume 406 of *Lecture Notes in Control and Information Sciences.* Springer, Berlin, 2010.

[Bit86] Sergio Bittanti. Deterministic and stochastic linear periodic systems. In Sergio Bittanti, editor, *Time Series and Linear Systems*, volume 86 of *Lecture Notes in Control and Information Sciences*, chapter 5, pages 141–182. Springer, 1986.

[Bla95] Franco Blanchini. Nonquadratic Lyapunov functions for robust control. *Automatica*, 31(3):451–461, 1995.

[BM99] Alberto Bemporad and Manfred Morari. Control of systems integrating logic, dynamics, and constraints. *Automatica*, 35(3):407–427, 1999.

[BMDP02] Alberto Bemporad, Manfred Morari, Vivek Dua, and Efstratios N. Pistikopoulos. The explicit linear quadratic regulator for constrained systems. *Automatica*, 38(1):3–20, 2002.

[Bor03] Francesco Borrelli. *Constrained Optimal Control of Linear and Hybrid Systems*, volume 290 of *Lecture Notes in Control and Information Sci-*

ences. Springer, Berlin, 2003.

[BPD93] Peter H. Bauer, Kamal Premaratne, and J. Durán. A necessary and sufficient condition for robust asymptotic stability of time-variant discrete systems. *IEEE Transactions on Automatic Control*, 38(9):1427–1430, 1993.

[Bra98] Michael S. Branicky. Multiple Lyapunov functions and other analysis tools for switched and hybrid systems. *IEEE Transactions on Automatic Control*, 43(4):475–482, 1998.

[BT96] Dimitri P. Bertsekas and John N. Tsitsiklis. *Neuro-Dynamic Programming.* Athena Scientific, Belmont, MA, 1996.

[Bus01] Linda G. Bushnell. Networks and control. *IEEE Control Systems Magazine*, 21(1):22–23, 2001.

[But05] Giorgio C. Buttazzo. *Hard Real-Time Computing Systems: Predictable Scheduling Algorithms and Applications.* Springer, New York, NY, 2nd edition, 2005.

[BV04] Stephen Boyd and Lieven Vandenberghe. *Convex Optimization.* Cambridge University Press, Cambridge, 2004.

[CA06] Anton Cervin and Peter Alriksson. Optimal on-line scheduling of multiple control tasks: A case study. In *Proceedings of the 18th Euromicro Conference on Real-Time Systems*, pages 141–150, 2006.

[CB04] Eduardo F. Camacho and Carlos Bordons. *Model Predictive Control.* Advanced Textbooks in Control and Signal Processing. Springer, London, 2nd edition, 2004.

[CGM02] Francesco A. Cuzzola, Jose C. Geromel, and Manfred Morari. An improved approach for constrained robust model predictive control. *Automatica*, 38(7):1183–1189, 2002.

[CH08] Anton Cervin and Toivo Henningsson. Scheduling of event-triggered controllers on a shared network. In *Proceedings of the 47th IEEE Conference on Decision and Control*, pages 3601–3606, 2008.

[CHL+03] Anton Cervin, Dan Henriksson, Bo Lincoln, Johan Eker, and Karl-Erik Årzén. How does control timing affect performance? *IEEE Control Systems Magazine*, 23(3):16–30, 2003.

[CK08] Ward Cheney and David Kincaid. *Numerical Mathematics and Computing.* Thomson Brooks/Cole, Belmont, CA, 6th edition, 2008.

[CM96] D. Chmielewski and V. Manousiouthakis. On constrained infinite-time linear quadratic optimal control. *Systems & Control Letters*, 29(11):121–129, 1996.

[CM02] Francesco A. Cuzzola and Manfred Morari. An LMI approach for H_∞ analysis and control of discrete-time piecewise affine systems. *International Journal of Control*, 75(16&17):1293–1301, 2002.

[CvdWHN09] Marieke B. G. Cloosterman, Nathan van de Wouw, W. P. M. H. Heemels, and Hendrik Nijmeijer. Stability of networked control systems with uncertain time-varying delays. *IEEE Transactions on Automatic Control*, 54(7):1575–1580, 2009.

[DBPL00] Raymond A. DeCarlo, Michael S. Branicky, Stefan Pettersson, and Bengt Lennartson. Perspectives and results on the stability and stabilizability of hybrid systems. *Proceedings of the IEEE*, 88(7):1069–1082, 2000.

[DRI02] Jamal Daafouz, Pierre Riedinger, and Claude Iung. Stability analysis and control synthesis for switched systems: A switched Lyapunov function approach. *IEEE Transactions on Automatic Control*, 47(11):1883–1887, 2002.

[EKSS00] Sebastian Engell, Stefan Kowalewski, Christian Schulz, and Olaf Stursberg. Continuous-discrete interactions in chemical processing plants. *Proceedings of the IEEE*, 88(7):1050–1068, 2000.

[FCMM02] Giancarlo Ferrari-Trecate, Francesco A. Cuzzola, Domenico Mignone, and Manfred Morari. Analysis of discrete-time piecewise affine and hybrid systems. *Automatica*, 38(12):2139–2146, 2002.

[FPW90] Gene F. Franklin, J. David Powell, and Michael L. Workman. *Digital Control of Dynamic Systems*. Addison-Wesley, Reading, MA, 2nd edition, 1990.

[GC06] Jose C. Geromel and Patrizio Colaneri. Stability and stabilization of discrete time switched systems. *International Journal of Control*, 79(7):719–728, 2006.

[GCHM06] Vijay Gupta, Timothy H. Chung, Babak Hassibi, and Richard M. Murray. On a stochastic sensor selection algorithm with applications in sensor scheduling and sensor coverage. *Automatica*, 42(2):251–260, 2006.

[GIL07] Daniel Görges, Michal Izák, and Steven Liu. Optimal control of systems with resource constraints. In *Proceedings of the 46th IEEE Conference on Decision and Control*, pages 1070–1075, 2007.

[GIL09] Daniel Görges, Michal Izák, and Steven Liu. Optimal control and scheduling of networked control systems. In *Proceedings of the 48th IEEE Conference on Decision and Control and 28th Chinese Control Conference*, pages 5839–5844, 2009.

[GIL11] Daniel Görges, Michal Izák, and Steven Liu. Optimal control and scheduling of switched systems. *IEEE Transactions on Automatic Control*, 56(1):135–140, 2011.

[GKBM05] Pascal Grieder, Michal Kvasnica, Mato Baotić, and Manfred Morari. Stabilizing low complexity feedback control of constrained piecewise affine systems. *Automatica*, 41(10):1683–1694, 2005.

[GLM03] Tobias Geyer, Mats Larsson, and Manfred Morari. Hybrid emergency voltage control in power systems. In *Proceedings of the European Control Conference 2003*, 2003.

[GMTT04] Gene Grimm, Michael J. Messina, Sezai E. Tuna, and Andrew R. Teel. Examples when nonlinear model predictive control is nonrobust. *Automatica*, 40(10):1729–1738, 2004.

[GMTT05] Gene Grimm, Michael J. Messina, Sezai E. Tuna, and Andrew R. Teel. Model predictive control: For want of a local control Lyapunov function, all is not lost. *IEEE Transactions on Automatic Control*, 50(5):546–558, 2005.

[GR08] Lars Grüne and Anders Rantzer. On the infinite horizon performance of receding horizon controllers. *IEEE Transactions on Automatic Control*, 53(9):2100–2111, 2008.

[Grü09] Lars Grüne. Analysis and design of unconstrained nonlinear MPC schemes for finite and infinite dimensional systems. *SIAM Journal on Control and Optimization*, 48(2):1206–1228, 2009.

[GST09] Rafal Goebel, Ricardo G. Sanfelice, and Andrew R. Teel. Hybrid dynamical systems: Robust stability and control for systems that combine continuous-time and discrete-time dynamics. *IEEE Control Systems Magazine*, 29(2):28–93, 2009.

[GV96] Gene H. Golub and Charles F. Van Loan. *Matrix Computations*. The Johns Hopkins University Press, Baltimore, MD, 3rd edition, 1996.

[Ham50] Richard W. Hamming. Error detecting and error correcting codes. *The Bell System Technical Journal*, 29(2):147–160, 1950.

[HC05] Dan Henriksson and Anton Cervin. Optimal on-line sampling period assignment for real-time control tasks based on plant state information. In *Proceedings of the 44th IEEE Conference on Decision and Control and European Control Conference 2005*, pages 4469–4474, 2005.

[HDB01] W. P. M. H. Heemels, Bart De Schutter, and Alberto Bemporad. Equivalence of hybrid dynamical models. *Automatica*, 37(7):1085–1091, 2001.

[HJ85] Roger A. Horn and Charles R. Johnson. *Matrix Analysis*. Cambridge University Press, Cambridge, 1985.

[HL05] Dimitrios Hristu-Varsakelis and William S. Levine, editors. *Handbook of Networked and Embedded Control Systems.* Birkhäuser, Boston, MA, 2005.

[HM99] João P. Hespanha and A. Stephen Morse. Stability of switched systems with average dwell-time. In *Proceedings of the 38th IEEE Conference on Decision and Control*, pages 2655–2660, 1999.

[HNX07] João P. Hespanha, Payam Naghshtabrizi, and Yonggang Xu. A survey of recent results in networked control systems. *Proceedings of the IEEE*, 95(1):138–162, 2007.

[Hri01] Dimitrios Hristu-Varsakelis. Feedback control systems as users of a shared network: Communication sequences that guarantee stability. In *Proceedings of 40th IEEE Conference on Decision and Control*, pages 3631–3636, 2001.

[Hri07] Dimitrios Hristu-Varsakelis. On the period of communication policies for networked control systems, and the question of zero-order holding. In *Proceedings of the 46th IEEE Conference on Decision and Control*, pages 38–43, 2007.

[IGL07] Michal Izák, Daniel Görges, and Steven Liu. On stability and control of systems with time-varying sampling period and time delay. In *Proceedings of the 7th IFAC Symposium on Nonlinear Control Systems*, pages 1056–1061, 2007.

[IGL08] Michal Izák, Daniel Görges, and Steven Liu. Stability and control of systems with uncertain time-varying sampling period and time delay. In *Proceedings of the 17th IFAC World Congress*, pages 11514–11519, 2008.

[IGL10a] Michal Izák, Daniel Görges, and Steven Liu. Optimal control of networked control systems with uncertain time-varying transmission delay. In *Proceedings of the 2nd IFAC Workshop on Distributed Estimation and Control in Networked Systems*, pages 13–18, 2010.

[IGL10b] Michal Izák, Daniel Görges, and Steven Liu. Stabilization of systems with variable and uncertain sampling period and time delay. *Nonlinear Analysis: Hybrid Systems*, 4(2):291–305, 2010.

[JG03] Tor A. Johansen and Alexandra Grancharova. Approximate explicit constrained linear model predictive control via orthogonal search tree. *IEEE Transactions on Automatic Control*, 48(5):810–815, 2003.

[JH05] Ali Jadbabaie and John Hauser. On the stability of receding horizon control with a general terminal cost. *IEEE Transactions on Automatic Control*, 50(5):674–678, 2005.

[JR98] Mikael Johansson and Anders Rantzer. Computation of piecewise quadratic Lyapunov functions for hybrid systems. *IEEE Transactions on Automatic Control*, 43(4):555–559, 1998.

[JTN05] Karl Henrik Johansson, Martin Törngren, and Lars Nielsen. Vehicle applications of controller area network. In Dimitrios Hristu-Varsakelis and William S. Levine, editors, *Handbook of Networked and Embedded Control Systems*. Birkhäuser, Boston, MA, 2005.

[JYH99] Ali Jadbabaie, Jie Yu, and John Hauser. Stabilizing receding horizon control of nonlinear systems: A control Lyapunov function approach. In *Proceedings of the 1999 American Control Conference*, pages 1535–1539, 1999.

[Kal60] R. E. Kalman. Contributions to the theory of optimal control. *Boletín de la Sociedad Matemática Mexicana*, 5:102–119, 1960.

[KBM96] Mayuresh V. Kothare, Venkataramanan Balakrishnan, and Manfred Morari. Robust constrained model predictive control using linear matrix inequalities. *Automatica*, 32(10):1361–1379, 1996.

[KG88] S. S. Keerthi and E. G. Gilbert. Optimal infinite-horizon feedback laws for a general class of constrained discrete-time systems: Stability and moving-horizon approximations. *Journal of Optimization Theory and Applications*, 57(2):265–293, 1988.

[KH05] Wook Hyun Kwon and Soohee Han. *Receding Horizon Control: Model Predictive Control for State Models*. Advanced Textbooks in Control and Signal Processing. Springer, London, 2005.

[KM02] Eric C. Kerrigan and David Q. Mayne. Optimal control of constrained, piecewise affine systems with bounded disturbances. In *Proceedings of the 41st IEEE Conference on Decision and Control*, pages 1552–1557, 2002.

[KS72] Huibert Kwakernaak and Raphael Sivan. *Linear Optimal Control Systems*. Wiley-Interscience, New York, NY, 1972.

[LA05] Hai Lin and Panos J. Antsaklis. Stability and persistent disturbance attenuation properties for a class of networked control systems: Switched system approach. *International Journal of Control*, 78(18):1447–1458, 2005.

[LA09] Hai Lin and Panos J. Antsaklis. Stability and stabilizability of switched linear systems: A survey of recent results. *IEEE Transactions on Automatic Control*, 54(2):308–322, 2009.

[Laz06] Mircea Lazar. *Model Predictive Control of Hybrid Systems: Stability and Robustness*. PhD thesis, Technische Universiteit Eindhoven, Eindhoven,

The Netherlands, 2006.

[LB02] Bo Lincoln and Bo Bernhardsson. LQR optimization of linear system switching. *IEEE Transactions on Automatic Control*, 47(10):1701–1705, 2002.

[LCRM04] Wilbur Langson, Ioannis Chryssochoos, Saša V. Raković, and David Q. Mayne. Robust model predictive control using tubes. *Automatica*, 40(1):125–133, 2004.

[Lee08] Edward A. Lee. Cyber physical systems: Design challenges. In *Proceedings of the 11th IEEE Symposium on Object/Component/Service-Oriented Real-Time Distributed Computing*, pages 363–369, 2008.

[Löf04] Johan Löfberg. YALMIP : A toolbox for modeling and optimization in MATLAB. In *Proceedings of the 2004 IEEE International Symposium on Computer Aided Control Systems Design*, pages 284–289, 2004.

[LGOH09] Junqi Liu, Azwirman Gusrialdi, Dragan Obradovic, and Sandra Hirche. Study on the effect of time delay on the performance of distributed power grids with networked cooperative control. In *Proceedings of the 1st IFAC Workshop on Distributed Estimation and Control in Networked Systems*, pages 168–173, 2009.

[LH09] Mircea Lazar and W. P. M. H. Heemels. Predictive control of hybrid systems: Input-to-state stability results for sub-optimal solutions. *Automatica*, 45(1):180–185, 2009.

[LHB09] Stefano Longo, Guido Herrmann, and Phil Barber. Optimization approaches for controller and schedule codesign in networked control. In *Proceedings of the 6th IFAC Symposium on Robust Control Design*, pages 301–306, 2009.

[LHT09] Mircea Lazar, W. P. M. H. Heemels, and Andrew R. Teel. Lyapunov functions, stability and input-to-state stability subtleties for discrete-time discontinuous systems. *IEEE Transactions on Automatic Control*, 54(10):2421–2425, 2009.

[LHWB06] Mircea Lazar, W. P. M. H. Heemels, Siep Weiland, and Alberto Bemporad. Stabilizing model predictive control of hybrid systems. *IEEE Transactions on Automatic Control*, 51(11):1813–1818, 2006.

[Lib03] Daniel Liberzon. *Switching in Systems and Control.* Birkhäuser, Boston, MA, 2003.

[LL09] Jan Lunze and Françoise Lamnabhi-Lagarrigue, editors. *Handbook of Hybrid Systems Control: Theory, Tools, Applications.* Cambridge University Press, Cambridge, 2009.

[LM99] Daniel Liberzon and A. Stephen Morse. Basic problems in stability and design of switched systems. *IEEE Control Systems Magazine*, 19(5):59–70, 1999.

[LMVF08] Camilo Lozoya, Pau Martí, Manel Velasco, and Josep M. Fuertes. Control performance evaluation of selected methods of feedback scheduling of real-time control tasks. In *Proceedings of the 17th IFAC World Congress*, pages 10668–10673, 2008.

[LR01] Bo Lincoln and Anders Rantzer. Optimizing linear system switching. In *Proceedings of the 40th IEEE Conference on Decision and Control*, pages 2063–2068, 2001.

[LR02] Bo Lincoln and Anders Rantzer. Suboptimal dynamic programming with error bounds. In *Proceedings of the 41st IEEE Conference on Decision and Control*, pages 2354–2359, 2002.

[LR06] Bo Lincoln and Anders Rantzer. Relaxing dynamic programming. *IEEE Transactions on Automatic Control*, 51(8):1249–1260, 2006.

[LTEP96] Bengt Lennartson, Michael Tittus, Bo Egardt, and Stefan Pettersson. Hybrid systems in process control. *IEEE Control Systems Magazine*, 16(5):45–56, 1996.

[LVM07] Camilo Lozoya, Manel Velasco, and Pau Martí. A 10-year taxonomy on prior work on sampling period selection for resource-constrained real-time control systems. In *Proceedings of the 19th Euromicro Conference on Real-Time Systems (Work-in-Progress Session)*, 2007.

[LZFA05] Hai Lin, Guisheng Zhai, Lei Fang, and Panos J. Antsaklis. Stability and $\mathcal{H}_\infty$ performance preserving scheduling policy for networked control systems. In *Proceedings of the 16th IFAC World Congress*, 2005.

[MÅB+03] Richard M. Murray, Karl J. Åström, Stephen P. Boyd, Roger W. Brockett, and Gunter Stein. Future directions in control in an information-rich world. *IEEE Control Systems Magazine*, 23(2):20–33, 2003.

[MAB+10] Sébastien Mariéthoz, Stefan Almér, Mihai Bâja, Andrea Giovanni Beccuti, Diego Patino, Andreas Wernrud, Jean Buisson, Hervé Cormerais, Tobias Geyer, Hisaya Fujioka, Ulf T. Jönsson, Chung-Yao Kao, Manfred Morari, Georgios Papafotiou, Anders Rantzer, and Pierre Riedinger. Comparison of hybrid control techniques for buck and boost DC-DC converters. *IEEE Transactions on Control Systems Technology*, 18(5):1126–1145, 2010.

[Mac02] Jan M. Maciejowski. *Predictive Control with Constraints*. Pearson Education, Harlow, 2002.

[Mao03] Wei-Jie Mao. Robust stabilization of uncertain time-varying discrete systems and comments on "An improved approach for constrained robust model predictive control". *Automatica*, 39(6):1109–1112, 2003.

[Mar03] Horacio J. Marquez. *Nonlinear Control Systems: Analysis and Design*. John Wiley & Sons, Hoboken, NJ, 2003.

[Mey00] Carl Dean Meyer. *Matrix Analysis and Applied Linear Algebra*. Society for Industrial and Applied Mathematics (SIAM), Philadelphia, PA, 2000.

[MFM00] Domenico Mignone, Giancarlo Ferrari-Trecate, and Manfred Morari. Stability and stabilization of piecewise affine and hybrid systems: An LMI approach. In *Proceedings of the 39th IEEE Conference on Decision and Control*, pages 504–509, 2000.

[ML99] Manfred Morari and Jay H. Lee. Model predictive control: Past, present and future. *Computers & Chemical Engineering*, 23(4-5):667–682, 1999.

[MM93] Hannah Michalska and David Q. Mayne. Robust receding horizon control of constrained nonlinear systems. *IEEE Transactions on Automatic Control*, 38(11):1623–1633, 1993.

[MP89] A. P. Molchanov and Ye. S. Pyatnitskiy. Criteria of asymptotic stability of differential and difference inclusions encountered in control theory. *Systems & Control Letters*, 13(1):59–64, 1989.

[MPSS99] A. Stephen Morse, Constantinos C. Pantelides, S. Shankar Sastry, and Hans Schumacher. Introduction to the special issue on hybrid systems. *Automatica*, 35(3):347–348, 1999.

[MR03] David Q. Mayne and Saša V. Raković. Model predictive control of constrained piecewise affine discrete-time systems. *International Journal of Robust and Nonlinear Control*, 13(3-4):261–279, 2003.

[MRRS00] David Q. Mayne, James B. Rawlings, Christopher V. Rao, and Pierre O. M. Scokaert. Constrained model predictive control: Stability and optimality. *Automatica*, 36(6):789–814, 2000.

[MT07] James R. Moyne and Dawn M. Tilbury. The emergence of industrial control networks for manufacturing control, diagnostics, and safety data. *Proceedings of the IEEE*, 95(1):29–47, 2007.

[MV78] Cleve Moler and Charles Van Loan. Nineteen dubious ways to compute the exponential of a matrix. *SIAM Review*, 20(4):801–836, 1978.

[NHL05] Thomas Nolte, Hans Hansson, and Lucia Lo Bellot. Automotive communications – Past, current and future. In *Proceedings of the 10th IEEE Conference on Emerging Technologies and Factory Automation*, pages 985–992, 2005.

[NP97] Vesna Nevistić and James A. Primbs. Finite receding horizon linear quadratic control: A unifying theory for stability and performance analysis. Technical Report CIT-CDS 97-001, California Institute of Technology, Control and Dynamical Systems, Pasadena, CA, USA, 1997.

[OFM07] Reza Olfati-Saber, J. Alex Fax, and Richard M. Murray. Consensus and cooperation in networked multi-agent systems. *Proceedings of the IEEE*, 95(1):215–233, 2007.

[Olf07] Reza Olfati-Saber. Distributed Kalman filtering for sensor networks. In *Proceedings of the 46th IEEE Conference on Decision and Control*, pages 5492–5498, 2007.

[Pet03] Stefan Pettersson. Synthesis of switched linear systems. In *Proceedings of the 42nd IEEE Conference on Decision and Control*, pages 5283–5288, 2003.

[PN00] James A. Primbs and Vesna Nevistić. Feasibility and stability of constrained finite receding horizon control. *Automatica*, 36(7):965–971, 2000.

[PND98] James A. Primbs, Vesna Nevistić, and John C. Doyle. On receding horizon extensions and control Lyapunov functions. In *Proceedings of the 1998 American Control Conference*, pages 3276–3280, 1998.

[Poo10] Radha Poovendran. Cyber-physical systems: Close encounters between two parallel worlds. *Proceedings of the IEEE*, 98(8):1363–1366, 2010.

[Pow07] Warren B. Powell. *Approximate dynamic programming: Solving the curses of dimensionality.* Wiley Series in Probability and Statistics. John Wiley & Sons, Hoboken, NJ, 2007.

[PP02] Athanasios Papoulis and S. Unnikrishna Pillai. *Probability, Random Variables, and Stochastic Processes.* McGraw-Hill, Boston, MA, 4th edition, 2002.

[QB03] S. Joe Qin and Thomas A. Badgwell. A survey of industrial model predictive control technology. *Control Engineering Practice*, 11(7):733–764, 2003.

[Ran06] Anders Rantzer. Relaxed dynamic programming in switching systems. *IEE Proceedings – Control Theory and Applications*, 153(5):567–574, 2006.

[RJ09] Maben Rabi and Karl Henrik Johansson. Scheduling packets for event-triggered control. In *Proceedings of the European Control Conference 2009*, pages 3779–3784, 2009.

[RM93] James B. Rawlings and Kenneth R. Muske. The stability of constrained receding horizon control. *IEEE Transactions on Automatic Control*,

38(10):1512–1516, 1993.

[RM09] James B. Rawlings and David Q. Mayne. *Model Predictive Control: Theory and Design.* Nob Hill Publishing, Madison, WI, 2009.

[RS00] Henrik Rehbinder and Martin Sanfridson. Integration of off-line scheduling and optimal control. In *Proceedings of the 12th Euromicro Conference on Real-Time Systems*, pages 137–143, 2000.

[RS04] Henrik Rehbinder and Martin Sanfridson. Scheduling of a limited communication channel for optimal control. *Automatica*, 40(3):491–500, 2004.

[San04] Martin Sanfridson. *Quality of Control and Real-Time Scheduling : Allowing for Time-Variations in Computer Control Systems.* PhD thesis, Royal Institute of Technology, Department of Machine Design, Stockholm, Sweden, 2004.

[Sch09] Luca Schenato. To zero or to hold control inputs with lossy links? *IEEE Transactions on Automatic Control*, 54(5):1093–1099, 2009.

[SG05a] Zhendong Sun and Shuzhi Sam Ge. Analysis and synthesis of switched linear control systems. *Automatica*, 41(2):181–195, 2005.

[SG05b] Zhendong Sun and Shuzhi Sam Ge. *Switched Linear Systems: Control and Design.* Communications and Control Engineering. Springer, 2005.

[SGIL10] Stefan Simon, Daniel Görges, Michal Izák, and Steven Liu. Periodic observer design for networked embedded control systems. In *Proceedings of the 2010 American Control Conference*, pages 4253–4258, 2010.

[SKMJ08] Jørgen Spjøtvold, Eric C. Kerrigan, David Q. Mayne, and Tor A. Johansen. Inf-sup control of discontinuous piecewise affine systems. *International Journal of Robust and Nonlinear Control*, 19(13):1471–1492, November 2008.

[SL91] Jean-Jacques E. Slotine and Weiping Li. *Applied Nonlinear Control.* Prentice-Hall, Englewood Cliffs, NJ, 1991.

[SLSS96] D. Seto, J.P. Lehoczky, L. Sha, and K.G. Shin. On task schedulability in real-time control systems. In *Proceedings of the 17th IEEE Real-Time Systems Symposium*, pages 13–21, 1996.

[SM98] Pierre O. M. Scokaert and David Q. Mayne. Min-max feedback model predictive control for constrained linear systems. *IEEE Transactions on Automatic Control*, 43(8):1136–1142, 1998.

[SMR99] Pierre O. M. Scokaert, David Q. Mayne, and James B. Rawlings. Suboptimal model predictive control (feasibility implies stability). *IEEE Transactions on Automatic Control*, 44(3):648–654, 1999.

[SP05] Sigurd Skogestad and Ian Postlethwaite. *Multivariable Feedback Control: Theory and Design.* Wiley, Chichester, 2nd edition, 2005.

[SR98] Pierre O. M. Scokaert and James B. Rawlings. Constrained linear quadratic regulation. *IEEE Transactions on Automatic Control*, 43(8):1163–1169, 1998.

[SSA10] Daniel Simon, Ye-Qiong Song, and Christophe Aubrun, editors. *Co-design Approaches to Dependable Networked Control Systems.* ISTE and Wiley, London and Hoboken, NJ, 2010.

[SSF+04] Bruno Sinopoli, Luca Schenato, Massimo Franceschetti, Kameshwar Poolla, Michael I. Jordan, and Shankar S. Sastry. Kalman filtering with intermittent observations. *IEEE Transactions on Automatic Control*, 49(9):1453–1464, 2004.

[SSF+07] Luca Schenato, Bruno Sinopoli, Massimo Franceschetti, Kameshwar Poolla, and S. Shankar Sastry. Foundations of control and estimation over lossy networks. *Proceedings of the IEEE*, 95(1):163–187, 2007.

[Stu99] Jos F. Sturm. Using SeDuMi 1.02, a MATLAB toolbox for optimization over symmetric cones. *Optimization Methods and Software*, 11&12(1-4):625–653, 1999.

[SWM+07] Robert Shorten, Fabian Wirth, Oliver Mason, Kai Wulff, and Christopher King. Stability criteria for switched and hybrid systems. *SIAM Review*, 49(4):545–592, 2007.

[Tab07] Paulo Tabuada. Event-triggered real-time scheduling of stabilizing control tasks. *IEEE Transactions on Automatic Control*, 52(9):1680–1685, 2007.

[TC03] Yodyium Tipsuwan and Mo-Yuen Chow. Control methodologies in networked control systems. *Control Engineering Practice*, 11(10):1099–1111, 2003.

[TJB03] Petter Tøndel, Tor A. Johansen, and Alberto Bemporad. Evaluation of piecewise affine control via binary search tree. *Automatica*, 39(5):945–950, 2003.

[TTT99] Kim-Chuan Toh, Michael J. Todd, and Reha H. Tütüncü. SDPT3 – A MATLAB software package for semidefinite programming, Version 1.3. *Optimization Methods and Software*, 11&12(1-4):545–581, 1999.

[TTT03] Reha H. Tütüncü, Kim-Chuan Toh, and Michael J. Todd. Solving semidefinite-quadratic-linear programs using SDPT3. *Mathematical Programming Series B*, 95(2):189–217, 2003.

[Van78] Charles F. Van Loan. Computing integrals involving the matrix exponential. *IEEE Transactions on Automatic Control*, 23(3):395–404, 1978.

[vdSS99] Arjan J. van der Schaft and Hans Schumacher. *An Introduction to Hybrid Dynamical Systems*, volume 251 of *Lecture Notes in Control and Information Sciences*. Springer, Berlin, 1999.

[Vid02] Mathukumalli Vidyasagar. *Nonlinear Systems Analysis*. Society for Industrial and Applied Mathematics (SIAM), Philadelphia, PA, 2nd edition, 2002.

[VZA+10] Michael P. Vitus, Wei Zhang, Alessandro Abate, Jianghai Hu, and Claire J. Tomlin. On efficient sensor scheduling for linear dynamical systems. In *Proceedings of the 2010 American Control Conference*, pages 4833–4838, 2010.

[Wer07] Andreas Wernrud. Strategies for computing switching feedback controllers. In *Proceedings of the 2007 American Control Conference*, pages 1389–1394, 2007.

[Wit66] Hans S. Witsenhausen. A class of hybrid-state continuous-time dynamic systems. *IEEE Transactions on Automatic Control*, 11(2):161–167, 1966.

[WL09] Xiaofeng Wang and Michael D. Lemmon. Self-triggered feedback control systems with finite-gain $\mathcal{L}_2$ stability. *IEEE Transactions on Automatic Control*, 54(3):452–467, 2009.

[Wol09] Wayne Wolf. Cyber-physical systems. *Computer*, 42(3):88–89, 2009.

[WY01] Gregory C. Walsh and Hong Ye. Scheduling of networked control systems. *IEEE Control Systems Magazine*, 21(1):57–65, 2001.

[XS06] Feng Xia and Youxian Sun. Control-scheduling codesign: A perspective on integrating control and computing. *Dynamics of Continuous, Discrete and Impulsive Systems, Series B: Applications & Algorithms*, 13:1352–1358, 2006. Special Issue on ICSCA'06.

[XS08] Feng Xia and Youxian Sun. *Control and Scheduling Codesign: Flexible Resource Management in Real-Time Control Systems*. Zhejiang University Press and Springer, Hangzhou and Berlin, 2008.

[Yak77] Vladimir A. Yakubovich. The S-procedure in nonlinear control theory. *Vestnik Leningrad University Mathematics*, 4(1):73–93, 1977.

[Yan06] T. C. Yang. Networked control system: A brief survey. *IEE Proceedings – Control Theory and Applications*, 153(4):403–412, 2006.

[YJ10] Bo Yang and Mikael Johansson. Distributed optimization and games: A tutorial overview. In Alberto Bemporad, Maurice Heemels, and Mikael Johansson, editors, *Networked Control Systems*, volume 406 of *Lecture Notes in Control and Information Sciences*, chapter 4, pages 109–148. Springer, Berlin, 2010.

[ZAH09] Wei Zhang, Alessandro Abate, and Jianghai Hu. Efficient suboptimal solutions of switched LQR problems. In *Proceedings of the 2009 American Control Conference*, pages 1084–1091, 2009.

[Zam08] Sandro Zampieri. Trends in networked control systems. In *Proceedings of the 17th IFAC World Congress*, pages 2886–2894, 2008.

[ZH08] Wei Zhang and Jianghai Hu. On optimal quadratic regulation for discrete-time switched linear systems. In *Proceedings of the 11th International Conference on Hybrid Systems: Computation and Control*, pages 584–597, 2008.

[ZHA09] Wei Zhang, Jianghai Hu, and Alessandro Abate. On the value functions of the discrete-time switched LQR problem. *IEEE Transactions on Automatic Control*, 54(11):2669–2674, 2009.

[ZHL10] Wei Zhang, Jianghai Hu, and Jianming Lian. Quadratic optimal control of switched linear stochastic systems. *Systems & Control Letters*, 59(11):736–744, 2010.

[ZJLS01] Jun Zhang, Karl Henrik Johansson, John Lygeros, and Shankar Sastry. Zeno hybrid systems. *International Journal of Robust and Nonlinear Control*, 11(5):435–451, 2001.

Zusammenfassung

Optimale Regelung geschalteter Systeme

mit Anwendung für vernetzte eingebettete Regelungssysteme

Regelungssysteme entwickeln sich zunehmend zu Cyber-Physical Systems. Wesentlich für Cyber-Physical Systems ist eine enge Interaktion zwischen Informationsverarbeitung, Informationsübertragung, Regelung und der physikalischen Umgebung. Viele technische Systeme basieren schon heute auf einer solchen Interaktion unter anderem Verkehrsmittel wie Kraftfahrzeuge, Schienenfahrzeuge oder Flugzeuge, industrielle Anlagen, beispielsweise Fertigungsanlagen und verfahrenstechnische Anlagen, sowie Infrastruktursysteme wie Energienetze oder „intelligente" Gebäude. Cyber-Physical Systems sind somit allgegenwärtig. Deren technologische, ökonomische und soziologische Bedeutung ist folglich außerordentlich hoch.

Obwohl Cyber-Physical Systems bereits praktische Anwendungen durchdringen, sind deren theoretische Grundlagen noch unvollständig. Methoden zur Modellierung, Analyse und Synthese von Cyber-Physical Systems sind erforderlich, welche die komplexe Interaktion zwischen Informationsverarbeitung, Informationsübertragung, Regelung und der physikalischen Umgebung holistisch berücksichtigen. Anders können Funktionalitäts-, Effizienz-, Zuverlässigkeits- und Sicherheitsanforderungen an Cyber-Physical Systems bei zunehmender Komplexität nicht erfüllt werden. Fachrichtungen wie Mathematik, Informatik, Kommunikationstechnik und Regelungstechnik haben sich bisher weitgehend unabhängig voneinander entwickelt. Diese Fachrichtungen müssen zusammengeführt werden, um theoretische Grundlagen für Cyber-Physical Systems zu schaffen.

Aus regelungstheoretischer und regelungstechnischer Sicht sind folgende Forschungsrichtungen von besonderer Bedeutung:

Einerseits müssen Methoden zur Modellierung, Analyse und Regelung von Systemen mit sowohl kontinuierlicher Dynamik, welche die physikalische Umgebung und die Regelung bestimmt, als auch diskreter Dynamik, welche die Informationsverarbeitung und Informationsübertragung bestimmt, erforscht werden. Solche Systeme werden als hybride Systeme bezeichnet. Hybride Systeme charakterisieren neben Cyber-Physical Systems viele andere technische Systeme, beispielsweise Kraftfahrzeuge mit Schaltgetriebe, Umrichter mit Dioden und Transistoren oder verfahrenstechnische Anlagen mit schaltenden Ventilen. Methoden für hybride Systeme sind somit von übergeordneter Bedeutung.

Andererseits müssen Methoden zur Modellierung, Analyse und Synthese von vernetzten eingebetteten Regelungssystemen entwickelt werden. Die Regler, Sensoren und Aktoren sind hierbei über ein Kommunikationsnetzwerk verbunden, die Regler ferner auf Prozessoren implementiert, welche in die Anwendung eingebettet sind. Vernetzte eingebettete Regelungssysteme bilden somit einen wesentlichen Grundbaustein für Cyber-Physical Systems. Charakteristisch für vernetzte eingebettete Regelungssysteme sind starke Beschränkungen der Rechen- und Kommunikationsressourcen. Eine Integration von Scheduling, das heißt der zeitlichen Allokation der Ressourcen, und Regelung ist daher von außerordentlicher Bedeutung, um die limitierten Rechen- und Kommunikationsressourcen effizient zu nutzen. Entsprechende Ansätze werden subsumiert als Regler-Scheduler-Codesign. Charakteristisch für vernetzte eingebettete Regelungssysteme sind ferner Effekte wie unsichere zeitvariante Abtastzeiten und Latenzen, Paketverluste und Quantisierungseffekte. Diese Effekte müssen sowohl bei der Stabilitätsanalyse als auch bei der Reglersynthese berücksichtigt werden, um die Stabilität und die Regelgüte zu gewährleisten. Entsprechende Ansätze werden subsumiert als implementierungsbewusste Regelung.

Diese Dissertation hat das Ziel, zum Wissensstand in beiden Forschungsrichtungen beizutragen.

Erstes Ziel der Dissertation ist die Entwicklung von Optimierungsstrategien für Regelung und Scheduling von geschalteten Systemen, einer wichtigen Klasse und Abstraktion von hybriden Systemen. Geschaltete Systeme bestehen aus einer Menge von Untersystemen und einem Schaltgesetz, welches die Umschaltung zwischen den Untersystemen definiert. Ziel sind Strategien für einen gemeinsamen Entwurf eines Regelgesetzes und eines Schaltgesetzes für ein geschaltetes System derart, dass eine Kostenfunktion minimiert wird. Das zugrunde liegende Optimierungsproblem weist eine exponentielle Komplexität auf. Die Reduktion dieser Komplexität ist eine wesentliche wissenschaftliche Herausforderung. Während die Stabilität und Stabilisierung geschalteter Systeme in den letzten Jahren intensiv erforscht wurden, wurden Optimierungsstrategien für Regelung und Scheduling von geschalteten Systemen bisher nur selten untersucht. Ziel dieser Dissertation ist, den Wissensstand diesbezüglich zu erweitern mit Fokus auf zeitdiskrete lineare geschaltete Systeme.

Zweites Ziel der Dissertation ist die Entwicklung von Optimierungsstrategien für Regelung und Scheduling von vernetzten eingebetteten Regelungssystemen. Es werden Konzepte für ein Regler-Scheduler-Codesign angestrebt, welche die Stabilität garantieren, Implementierungseffekte berücksichtigen, ein Offline- und Online-Scheduling ermöglichen und eine Balancierung zwischen der Komplexität (offline und online) und der Regelgüte gestatten. Existierende Konzepte sind meist auf einzelne dieser Aspekte ausgerichtet.

Die Dissertation ist entsprechend in zwei Teile gegliedert.

Teil I konzentriert sich auf Optimierungsstrategien für Regelung und Scheduling von geschalteten Systemen.

Kapitel 1 enthält einen Überblick zu hybriden Systemen, geschalteten Systemen, der Stabilität und Stabilisierung von geschalteten Systemen sowie der optimalen Regelung von geschalteten Systemen. Grundlegende Konzepte werden eingeführt, der Stand der Forschung diskutiert sowie der Beitrag dieser Dissertation beschrieben.

Kapitel 2 ist fokussiert auf Optimierungsstrategien für Regelung und Scheduling von geschalteten Systemen mit endlichem Zeithorizont. Das Optimierungsproblem wird zunächst formalisiert, die Lösung wird dann basierend auf dynamischer Programmierung bestimmt, jedoch unter exponentieller Komplexität. Zur Komplexitätsreduktion wird ein notwendiges und hinreichendes Wegschneidekriterium formuliert. Das Wegschneidekriterium kann nicht analytisch, sondern nur durch Abrasterung des Zustandsraums ausgewertet werden. Es werden daher zwei hinreichende Wegscheidekriterien basierend unter anderem auf der S-Prozedur hergeleitet und Algorithmen zur analytischen Auswertung angegeben. Die Lösung des Optimierungsproblems wird ferner basierend auf relaxierter dynamischer Programmierung bestimmt. Eine Komplexitätsreduktion wird hierbei durch Relaxation der Optimalität innerhalb vorgegebener Schranken erzielt. Die Eigenschaften der Methoden werden anhand numerischer Beispiele illustriert, deren Effektivität anhand einer numerischen Studie.

Kapitel 3 ist fokussiert auf Optimierungsstrategien für Regelung und Scheduling von geschalteten Systemen mit bewegtem Zeithorizont. Hierfür werden Methoden der modellbasierten prädiktiven Regelung verwendet. Das Optimierungsproblem wird zunächst formalisiert, die Lösung wird dann ausgehend von den in Kapitel 2 eingeführten Methoden ermittelt. Es wird gezeigt, dass die Lösung explizit dargestellt werden kann als stückweise lineare Zustandsrückführung, welche auf Regionen des Zustandsraums definiert ist, die durch quadratische Formen charakterisiert sind. Die zugehörigen geschalteten Rückführmatrizen können offline berechnet und gespeichert werden. Die Berechnungen online werden dadurch reduziert auf die Auswertung der stückweise linearen Zustandsrückführung basierend auf einem Regionenzugehörigkeitstest. Die Stabilität wird durch die stückweise lineare Zustandsrückführung nicht inhärent garantiert. Es wird daher eine A-Priori-Stabilitätsbedingung basierend auf einem stabilisierenden Endkostenterm hergeleitet. Diese Stabilitätsbedingung ist anwendbar, wenn die Lösung des Optimierungsproblems durch dynamische Programmierung unter Wegschneiden, also die optimale Lösung, bestimmt wurde. Es werden ferner zwei A-Posteriori-Stabilitätskriterien vorgeschlagen, welche auch dann anwendbar sind, wenn die Lösung des Optimierungsproblems durch relaxierte dynamische Programmierung, also eine suboptimale Lösung, bestimmt wurde. Das erste A-Posteriori-Stabilitätskriterium basiert auf der Konstruktion einer stückweise quadratischen Ljapunow-Funktion. Ein wesentlicher Beitrag ist hierbei eine Methode zur Berücksichtigung der Regionen in der stückweise quadratischen Ljapunow-Funktion. Das zweite A-Posteriori-Stabilitätskriterium basiert auf der Verwendung der Wertefunktion aus der dynamischen Programmierung als potentielle Ljapunow-Funktion. Ferner wird ein Kriterium zur Analyse der Erreichbarkeit zwischen den Regionen angegeben. Die

Eigenschaften der Methoden werden durch numerische Beispiele illustriert, deren Effektivität durch numerische Studien.

Kapitel 4 ist fokussiert auf Optimierungsstrategien für Regelung und Scheduling von geschalteten Systemen mit unendlichem Zeithorizont und periodischem Schaltgesetz. Es werden sowohl ein Offline-Scheduling als auch ein Online-Scheduling betrachtet. Die zugehörigen Optimierungsprobleme werden zunächst formalisiert, deren Lösungen werden dann basierend auf periodischer Regelungstheorie und erschöpfender Suche bestimmt. Es wird gezeigt, dass die Lösung sowohl für Offline-Scheduling als auch für Online-Scheduling die Stabilität inhärent garantiert und dass die Lösung für Online-Scheduling explizit dargestellt werden kann als stückweise lineare Zustandsrückführung, welche auf durch quadratische Formen charakterisierten Regionen des Zustandsraums definiert ist. Die Komplexität des Optimierungsproblems nimmt exponentiell mit der Periode des Schaltgesetzes zu. Hierdurch ist insbesondere auch die Komplexität für die Auswertung der stückweise linearen Zustandsrückführung bestimmt, welche online erfolgt. Es werden daher drei Methoden zur Reduktion dieser Komplexität basierend auf Relaxation und Heuristiken vorgeschlagen. Bemerkenswerterweise führt ein Online-Scheduling stets zu gleichen oder geringeren Kosten als ein Offline-Scheduling. Die Effektivität der Methoden wird anhand umfangreicher numerischer Studien gezeigt.

Kapitel 5 beinhaltet eine Zusammenfassung der Methoden und Ergebnisse aus Teil I. Zusätzlich werden Anregungen für weitere Untersuchungen gegeben.

Teil II konzentriert sich auf Optimierungsstrategien für Regelung und Scheduling von vernetzten eingebetteten Regelungssystemen.

Kapitel 6 enthält einen Überblick zu vernetzten eingebetteten Regelungssystemen, zur implementierungsbewussten Regelung sowie zum Regler-Scheduler-Codesign. Es werden insbesondere Merkmale zur Klassifizierung von Konzepten für das Regler-Scheduler-Codesign erarbeitet. Existierende Konzepte werden anhand dieser Merkmale klassifiziert. Basierend hierauf werden der Stand der Forschung und aktuelle Tendenzen aufgezeigt sowie der Beitrag der Dissertation beschrieben.

Kapitel 7 umfasst die Modellierung vernetzter eingebetteter Regelungssysteme. Es wird eine Architektur bestehend aus mehreren Strecken eingeführt, welche über ein einzelnes Kommunikationsnetzwerk und mit einem einzelnen Prozessor geregelt werden. Eine solche Architektur ist generisch für vernetzte eingebettete Regelungssysteme mit geteilten Rechen- und Kommunikationsressourcen und auch häufig in praktischen Anwendungen anzutreffen. Für diese Architektur wird ein Rechen- und Kommunikationsmodell vorgestellt. Basierend hierauf werden ein dynamisches Modell und eine Kostenfunktion hergeleitet. Jede Strecke inklusive Sensor und Aktor wird durch eine zeitkontinuierliche lineare zeitinvariante Zustandsgleichung mit einer Eingangstotzeit zur Repräsentation von Rechen- und Übertragungszeiten (Latenzen) modelliert. Die zeitkontinuierliche Zustandsgleichung wird diskretisiert unter Verwendung eines speziellen Haltegliedes: Falls der Regelalgorithmus zu einer Strecke im Diskretisierungsintervall ausgeführt wird, wird eine neue Stellgröße innerhalb des Diskretisierungsintervalls berechnet, übertragen und aufgeschaltet. Andernfalls wird die alte Stellgröße über das Diskretisierungsintervall gehalten. Nach der Diskretisierung wird jede Strecke durch eine zeitdiskrete lineare zeitvariante Zustandsgleichung modelliert. Die zeitdiskreten Zustandsgleichungen der einzelnen Strecken werden zu einer blockdiagonalen zeitdiskreten Zustandsgleichung zusammengefasst, welche das gesamte vernetzte eingebettete Regelungssystem beschreibt. Analog wird hinsichtlich der Kostenfunktion vorgegangen. Mittels Diskretisierung unter Verwendung des speziellen Haltegliedes können Implementierungseffekte wie zeitvariante Abtastzeiten und Latenzen exakt in dem Modell und der Kostenfunktion abgebildet werden. Hierdurch wird insbesondere ein Vergleich verschiedener Schedules ermöglicht.

Kapitel 8 ist fokussiert auf das Regler-Scheduler-Codesign für vernetzte eingebettete Regelungssysteme. Das Optimierungsproblem, welches dem Regler-Scheduler-Codesign zugrunde liegt, wird zunächst formalisiert. Es wird dann gezeigt, dass die blockdiagonale zeitdiskrete Zustandsgleichung einem blockdiagonalen zeitdiskreten linearen geschalteten System entspricht. Die in Teil I eingeführten Methoden können somit zur Lösung des Optimierungsproblems herangezogen werden. Diese Methoden garantieren grundsätzlich die Stabilität und erlauben eine systematische Balancierung zwischen der Komplexität (offline und online) und der Regelgüte. Allerdings sind teils Erweiterungen erforderlich aufgrund der strukturellen Eigenschaften des blockdiagonalen geschalteten Systems.

Kapitel 9 ist fokussiert auf Optimierungsstrategien für Regelung und Scheduling von geschalteten Systemen mit bewegtem Zeithorizont über N' Schritte, einer Erweiterung der Optimierungsstrategien aus Kapitel 3 für vernetzte eingebettete Regelungssysteme. Die Lösung durch relaxierte dynamische Programmierung, die explizite Lösung, die A-Posteriori-Stabilitätsanalyse basierend auf einer stückweise quadratischen Ljapunow-Funktion und die Erreichbarkeitsanalyse werden hierzu unter Beachtung der strukturellen Eigenschaften des blockdiagonalen geschalteten Systems umfassend überarbeitet. Hierbei wird insbesondere die Lifting-Prozedur verwendet. Ferner wird eine Methode zur Suboptimalitätsanalyse eingeführt.

Kapitel 10 enthält eine Simulationsstudie zur vernetzten Regelung dreier invertierter Pendel. Die Eigenschaften der Methoden aus Teil I und II werden hierbei herausgestellt und verglichen. Die Ergebnisse unterstreichen, dass durch ein Regler-Scheduler-Codesign die Regelgüte eines vernetzten eingebetteten Regelungssystems im Vergleich zu Methoden, die in der industriellen Praxis angewendet werden, signifikant gesteigert werden kann. Ferner wird deutlich, dass ein Online-Scheduling allgemein zu einer erheblich höheren Regelgüte führt als ein Offline-Scheduling.

Kapitel 11 beinhaltet eine Zusammenfassung der Methoden und Ergebnisse aus Teil II sowie Vorschläge für zukünftige Untersuchungen.

Curriculum Vitae

Personal Data

Name	Daniel Görges
Born	7 January 1980 in Koblenz (Germany)
Address	Buchenlochstraße 66 B (Apt. 22) 67663 Kaiserslautern (Germany)
E-Mail	goerges@eit.uni-kl.de

Education

10/2000 – 12/2005	**University of Kaiserslautern**
	Subject: Informationstechnik (Information Technology) Specialization: Automatisierungstechnik (Automation)
	Degree: Dipl.-Ing. (equivalent to M. Sc.)
	Thesis: Aktive Dämpfung der Mastschwingungen von Regalbediengeräten (Active Vibration Control of Storage and Retrieval Machines / Kreissparkassenstiftung Award)
08/1990 – 06/1999	**Wilhelm-Hofmann-Gymnasium St. Goarshausen**
	Degree: Abitur
08/1986 – 06/1990	**Grundschule Halsenbach**

Professional Experience

01/2006 –	**University of Kaiserslautern**
	Department of Electrical and Computer Engineering Institute of Control Systems
	Research Associate (01/2006 – 04/2009 as Scholarship Holder)
08/2000 – 09/2000	**BOMAG GmbH Boppard**
	Internship (Mechanical Manufacturing Methods)

In der Reihe *„Forschungsberichte aus dem Lehrstuhl für Regelungssysteme"*, herausgegeben von Steven Liu, sind bisher erschienen:

1	Daniel Zirkel	Flachheitsbasierter Entwurf von Mehrgrößenregelungen am Beispiel eines Brennstoffzellensystems ISBN 978-3-8325-2549-1, 2010, 159 S. 35.00 €
2	Martin Pieschel	Frequenzselektive Aktivfilterung von Stromoberschwingungen mit einer erweiterten modellbasierten Prädiktivregelung ISBN 978-3-8325-2765-5, 2010, 160 S. 35.00 €
3	Philipp Münch	Konzeption und Entwurf integrierter Regelungen für Modulare Multilevel Umrichter ISBN 978-3-8325-2903-1, 2011, 183 S. 44.00 €
4	Jens Kroneis	Model-based trajectory tracking control of a planar parallel robot with redundancies ISBN 978-3-8325-2919-2, 2011, 279 S. 39.50 €
5	Daniel Görges	Optimal Control of Switched Systems with Application to Networked Embedded Control Systems ISBN 978-3-8325-3096-9, 2012, 201 S. 36.50 €

Alle erschienenen Bücher können unter der angegebenen ISBN im Buchhandel oder direkt beim Logos Verlag Berlin (www.logos-verlag.de, Fax: 030 - 42 85 10 92) bestellt werden.